T0140215

Domain Adaptation for Visual Understanding

Richa Singh · Mayank Vatsa · Vishal M. Patel ·
Nalini Ratha
Editors

Domain Adaptation
for Visual Understanding

 Springer

Editors
Richa Singh🆔
Indraprastha Institute of Information
Technology Delhi
New Delhi, India

Mayank Vatsa🆔
Indraprastha Institute of Information
Technology Delhi
New Delhi, India

Vishal M. Patel
Johns Hopkins University
Baltimore, MD, USA

Nalini Ratha🆔
IBM Thomas J. Watson Research Center
Yorktown Heights, NY, USA

ISBN 978-3-030-30673-1   ISBN 978-3-030-30671-7   (eBook)
https://doi.org/10.1007/978-3-030-30671-7

This Springer imprint is published by the registered company Springer Nature Switzerland AG
The registered company address is: Gewerbestrasse 11, 6330 Cham, Switzerland

Preface

In many real-world vision applications, there are very few or even no labeled samples, while an unrelated general domain is often available with a large number of labeled examples. For example, ImageNet contains millions of loosely labeled images over a large number of general classes of objects. On the other hand, a medical researcher may be interested in retrieving brain cancer fMRI scans closer to the patient's brain scan image. Such data may not be available in large volumes or may be expensive to put forth the effort to annotate their collections by themselves. The problem of a lack of training samples can be challenging because of the significant statistical distribution difference between the feature distributions of training samples from the known available domain and the application domain. Researchers have often resorted to many techniques such as fine-tuning, hard mining, transfer learning, and domain adaptation to effectively use the large training samples from one domain and still get benefits in the targeted application domain to explicitly handle variations in feature distributions.

In this edited volume, we address various challenges of domain adaptation for visual understanding through the following chapters. We start with a review of available methods in domain adaptation in general. The second chapter, by Issam H. Laradji and Reza Babanezhad, describes domain adaptation with deep metric learning. In the area of image-to-image translation, a novel unsupervised method called XGAN is presented in the third chapter, by Amélie Royer, Konstantinos Bousmalis, Stephan Gouws, Fred Bertsch, Inbar Mosseri, Forrester Cole, and Kevin Murphy. In the area of improving transferability of deep neural networks, Parijat Dube, Bishwaranjan Bhattacharjee, Elisabeth Petit-Bois, and Matthew Hill present an interesting way to find out which dataset in the base training is useful in target results, in the fourth chapter. Xinyan Yu, Ya Zhang, and Rui Zhang present an interesting idea in the fifth chapter, for when the target domain is in the area of video retrieval. In order to estimate the discrepancy between the source and the target domain, Mohammad Mahfujur Rahman, Clinton Fookes, Mahsa Baktashmotlagh, and Sridha Sridharan present a quantitative method in the sixth chapter. In order to enhance facial action recognition, Nishant Sankaran, Deen Dayal Mohan, Nagashri N. Lakshminarayana, Srirangaraj Setlur, and Venu Govindaraju present their ideas in the seventh chapter. An interesting concept of learning intuition, generalizing the idea of domain adaptation, is covered in eighth chapter, by Anush Sankaran, Mayank Vatsa, and Richa Singh. In the final chapter, by Xiyu Kong, Qiping Zhou, Yunyu Lai, Muming Zhao, and Chongyang Zhang, a novel interpolation-based tracking model is presented to address the tracking model degradation problem of existing CF-based methods.

As can be seen, these articles cover a wide range of domain adaptation topics for various vision applications such as object recognition, face recognition, and action and event recognition. We hope the diversity of the topics help the readers understand the challenges in domain adaptation and related problems.

New Delhi, India   Richa Singh
New Delhi, India   Mayank Vatsa
Baltimore, USA   Vishal M. Patel
Yorktown Heights, USA   Nalini Ratha

Contents

Domain Adaptation for Visual Understanding 1
Soumyadeep Ghosh, Richa Singh, Mayank Vatsa, Nalini Ratha
and Vishal M. Patel

**M-ADDA: Unsupervised Domain Adaptation with Deep Metric
Learning** 17
Issam H. Laradji and Reza Babanezhad

**XGAN: Unsupervised Image-to-Image Translation for Many-to-Many
Mappings** 33
Amélie Royer, Konstantinos Bousmalis, Stephan Gouws, Fred Bertsch,
Inbar Mosseri, Forrester Cole and Kevin Murphy

Improving Transferability of Deep Neural Networks 51
Parijat Dube, Bishwaranjan Bhattacharjee, Elisabeth Petit-Bois
and Matthew Hill

Cross-Modality Video Segment Retrieval with Ensemble Learning 65
Xinyan Yu, Ya Zhang and Rui Zhang

**On Minimum Discrepancy Estimation for Deep Domain
Adaptation** 81
Mohammad Mahfujur Rahman, Clinton Fookes, Mahsa Baktashmotlagh
and Sridha Sridharan

**Multi-modal Conditional Feature Enhancement for Facial Action
Unit Recognition** 95
Nagashri N. Lakshminarayana, Deen Dayal Mohan, Nishant Sankaran,
Srirangaraj Setlur and Venu Govindaraju

Intuition Learning 111
Anush Sankaran, Mayank Vatsa and Richa Singh

**Alleviating Tracking Model Degradation Using Interpolation-Based
Progressive Updating** 129
Xiyu Kong, Qiping Zhou, Yunyu Lai, Muming Zhao
and Chongyang Zhang

Index 143

Contributors

Reza Babanezhad Department of Computer Science, University of British Columbia, Vancouver, BC, Canada

Mahsa Baktashmotlagh Image and Video Laboratory, Queensland University of Technology (QUT), Brisbane, QLD, Australia

Fred Bertsch Google Brain, Mountain View, CA, USA

Bishwaranjan Bhattacharjee IBM Research AI, Yorktown Heights, NY, USA

Konstantinos Bousmalis Google Brain, London, UK;
Deepmind, London, UK

Forrester Cole Google Research, Cambridge, MA, USA

Parijat Dube IBM Research AI, Yorktown Heights, NY, USA

Clinton Fookes Image and Video Laboratory, Queensland University of Technology (QUT), Brisbane, QLD, Australia

Soumyadeep Ghosh Department of Computer Science and Engineering, IIIT Delhi, New Delhi, India

Stephan Gouws Google Brain, London, UK

Venu Govindaraju University at Buffalo, Buffalo, NY, USA

Matthew Hill IBM Research AI, Yorktown Heights, NY, USA

Xiyu Kong Institute of Image Communication and Network Engineering, Shanghai Jiao Tong University, Shanghai, China

Yunyu Lai State Grid Jiangxi Power Co. Ltd, Maintenance Branch, Nanchang, China

Nagashri N. Lakshminarayana University at Buffalo, Buffalo, NY, USA

Issam H. Laradji Department of Computer Science, University of British Columbia, Vancouver, BC, Canada

Deen Dayal Mohan University at Buffalo, Buffalo, NY, USA

Inbar Mosseri Google Research, Cambridge, MA, USA

Kevin Murphy Google Research, Mountain View, CA, USA

Vishal M. Patel Department of Electrical and Computer Engineering, Johns Hopkins University, Baltimore, MD, USA

Elisabeth Petit-Bois IBM Research AI, Yorktown Heights, NY, USA; Kennesaw State University, Kennesaw, GA, USA

Mohammad Mahfujur Rahman Image and Video Laboratory, Queensland University of Technology (QUT), Brisbane, QLD, Australia

Nalini Ratha IBM TJ Watson Research Center, Yorktown, NY, USA

Amélie Royer IST Austria, Klosterneuburg, Austria

Anush Sankaran IBM Research, Bangalore, India

Nishant Sankaran University at Buffalo, Buffalo, NY, USA

Srirangaraj Setlur University at Buffalo, Buffalo, NY, USA

Richa Singh Department of Computer Science and Engineering, IIIT Delhi, New Delhi, India

Sridha Sridharan Image and Video Laboratory, Queensland University of Technology (QUT), Brisbane, QLD, Australia

Mayank Vatsa Department of Computer Science and Engineering, IIIT Delhi, New Delhi, India

Xinyan Yu Cooperative Medianet Innovation Center, Shanghai Jiao Tong University, Minhang, China

Chongyang Zhang Institute of Image Communication and Network Engineering, Shanghai Jiao Tong University, Shanghai, China

Rui Zhang Cooperative Medianet Innovation Center, Shanghai Jiao Tong University, Minhang, China

Ya Zhang Cooperative Medianet Innovation Center, Shanghai Jiao Tong University, Minhang, China

Muming Zhao Institute of Image Communication and Network Engineering, Shanghai Jiao Tong University, Shanghai, China

Qiping Zhou State Grid Jiangxi Power Co. Ltd, Maintenance Branch, Nanchang, China

Domain Adaptation for Visual Understanding

Soumyadeep Ghosh, Richa Singh, Mayank Vatsa, Nalini Ratha
and Vishal M. Patel

Abstract Advances in visual understanding in the last two decades have been aided
by exemplary progress in machine learning and deep learning methods. One of the
principal issues of modern classifiers is generalization toward unseen testing data
which may have a distribution different to that of the training set. Further, classifiers
need to be adapted to scenarios where training data is made available online. Domain
adaptation based machine learning algorithms cater to these specific scenarios where
the classifiers are updated for inclusivity and generalizability. Such methods need to
encompass the covariate shift so that the trained model gives appreciable performance
on the testing data. In this chapter, we categorize, illustrate, and analyze different
domain adaptation based machine learning algorithms for visual understanding.

Keywords Domain adaptation · Deep learning · Image classification

S. Ghosh · R. Singh (✉) · M. Vatsa
Department of Computer Science and Engineering, IIIT Delhi, Okhla Industrial Estate, Phase III,
New Delhi 110020, India
e-mail: rsingh@iiitd.ac.in

S. Ghosh
e-mail: soumyadeepg@iiitd.ac.in

M. Vatsa
e-mail: mayank@iiitd.ac.in

N. Ratha
IBM TJ Watson Research Center, 1101 Kitchawan Road,
Yorktown, NY 10598, USA
e-mail: ratha@us.ibm.com

V. M. Patel
Department of Electrical and Computer Engineering, Johns Hopkins University,
3400 N Charles St, Baltimore, MD 21218-2625, USA
e-mail: vpatel36@jhu.edu

© Springer Nature Switzerland AG 2020
R. Singh et al. (eds.), *Domain Adaptation for Visual Understanding*,
https://doi.org/10.1007/978-3-030-30671-7_1

1

1 Introduction

In the last two decades, visual understanding has been at the forefront of machine learning applications. Several applications of computer vision and visual understanding have been explored such as face recognition [1–7] and segmentation [8], object tracking [9], scene understanding [10], gesture recognition [11], shape reconstruction and understanding [12], multimodal representation learning [8], image retrieval [13], and activity recognition [14]. The advancements in the applications of visual understanding have been possible due to the availability of training data, computing resources, and development of modern deep learning algorithms.

One of the fundamental assumptions of any pattern recognition algorithm is that the training and testing data are drawn from the same distribution. Considering the explosive increase of data available on the Internet (through websites such as Youtube and Flickr), it is not difficult to collect a large amount of data to train a system for visual understanding. However, such a multisource pool of data may result in the variation of training and testing distribution for the model being trained. To take an example, for object detection, a model is trained to detect objects in an indoor setting while testing can be done in outdoor scenarios. Due to the variation of background and illumination, testing performance can be poor using a model which is trained on data acquired in indoor settings. To illustrate with another example, in order to train a model for face recognition, the data on which training is performed is expected to be similar (in terms of a scenario of image acquisition) to the data on which the model would be evaluated. In this context, the type of data can be defined in terms of the covariates or variation in the modality/type of data. For face recognition, covariates refer to the resolution [15, 16], spectrum of image capture (visible/NIR/thermal) [17, 18], pose, illumination [19], expression, age [20], and disguise [21, 22]. As shown in Fig. 1, all variations in data are due to the fact that images can be captured under different acquisition scenarios. Therefore, any variation in the distribution of the training and testing sets would have an impact on the testing performance for a model [23–27]. In addition to this, the bias in a learning algorithm is also a related phenomenon. For instance, a model trained on faces of adults could give very different results when tested on the face images of babies [28, 29]. This observation on models trained in a different distribution to that of the testing data is seen across all applications of visual understanding.

The outline of this chapter is as follows. In Sect. 2, we start with a brief discussion on transfer learning and discuss its different types in Sect. 3. We illustrate domain adaptation and its applications in Sect. 4. Paradigms which are similar or related to domain adaptation are discussed in Sect. 4.3, followed by a summary and future research directions in Sect. 5.

(a) **(b)** **(c)** **(d)**

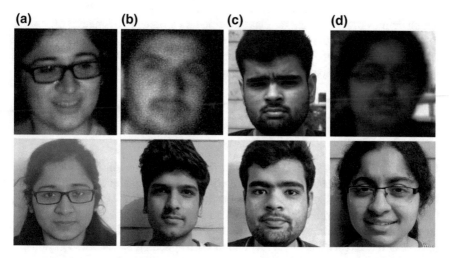

Fig. 1 The domain difference in training and testing sets. Upper row **a**: High-resolution NIR image, **b**: Low-resolution NIR image, **c**: High-resolution visible spectrum (VIS) image, and **d**: Low-resolution VIS image with poor illumination. Lower row **a–d**: Corresponding images of the same subject/class in controlled scenarios where the model is trained. The upper row shows all the different modalities of the test data, which is very different from the training data, which leads to unsatisfactory testing performance

2 Background: Transfer Learning

Any conventional machine learning algorithm [30–32] is designed with a basic assumption that the training and testing data belong to the same distribution. However, in several cases, not enough data is available such that the model can be trained. In such cases, the drift in training and testing sets can arise because of variations in domains, feature, or new task. A broad set of techniques that allow the model learned on a different domain or for a different task, to be adapted to the new domain/task falls under the purview of transfer learning.

2.1 Notations and Definitions

In the context of transfer learning, we define two primary aspects of a classification paradigm, namely, domain and task. The data on which training is performed is known as the source domain D_s and that of testing is known as the target domain D_t. A domain consists of two components, namely, a feature/sample space X and the marginal distribution $P(X)$. The source domain can be represented as $X_s = \{x_1^s, x_2^s \ldots x_n^s\}$ and the target domain is $X_t = \{x_1^t, x_2^t \ldots x_m^t\}$ on which the model would be evaluated. The source domain is thus given by $D_s = \{X_s, P(X_s)\}$ and the target domain by $D_t = \{X_t, P(X_t)\}$. The task τ is given by the labels Y of the data, which defines the categories for classification and a predictive function $g(.)$ which is to be learned from the training data. Thus the source task $\tau_s = \{Y_s, g(.)\}$, and the target task $\tau_t = \{Y_t, g(.)\}$ may be same or different. In cases where the domains are different, the feature spaces and/or the marginal distributions may be different.

2.2 Transfer Learning

In a particular setting, given a source domain D_s, target domain D_t, source task τ_s and target task τ_t, transfer learning represents the set of techniques which is used to learn an effective predictive function $g(.)$, where one or both of $D_s \neq D_t$ and $\tau_s \neq \tau_t$ holds true. While learning this predictive function, we may utilize all of the source domain data X_s and the source domain label space Y_s. In addition to this, we may have very few target domain data X_t and its corresponding label space Y_t available, which might help to adapt the function g_s which was learnt using D_s and τ_s alone.

Approaches for Transfer Learning This domain difference between the training and testing data is popularly known as domain or covariate shift [33] in related literature. This domain shift is relevant in real-life deployment settings for visual applications. A product developed for a certain scenario, if used under different settings with respect to the domain of data may produce unsatisfactory results. Let us take the example of a model trained on the source domain data D_s, and tested on the target domain D_t. In such a case, there are two approaches to improve upon the model trained using D_s. The first is to retrain the entire model using data which has a distribution similar to D_t. The second is to adapt the model so that it gives satisfactory performance on D_t as well. The former option might not be viable under situations where data pertaining to D_t is scarce. The second option has been heavily investigated by domain adaptation methods. Some of these methods may involve a small set of training data from D_t, which can be utilized to update the model trained using D_s. In domain adaptation, as discussed later which is a special case of transfer learning, the source and the target task remain the same but the data distribution of the train and the test sets differ (Fig. 1).

3 Categories of Transfer Learning

Depending on the availability and suitability of D_s, D_t, τ_s, and τ_t, we can classify transfer learning approaches into several categories (Fig. 2). Each of these categories cater to a particular train–test scenario for classification.

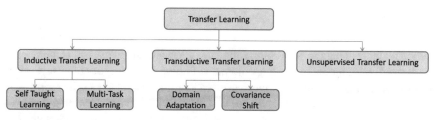

Fig. 2 Classification of transfer learning techniques (adapted from [34])

3.1 Inductive Transfer Learning

Scenarios where the source and the target tasks are different, fall under the purview of inductive transfer learning. The difference in the tasks may be attributed to a different label space (different categories/classes in train and test sets), or a different conditional probability distribution $P(y_i/x_i)$ where $y_i \in Y_t$ and $x_i \in X_t$. This conditional probability distribution is attributed to the distribution of the supervised samples, and depends on the amount of labeled data for each class. Zero-shot learning [35–37] is one of the most prominent examples for inductive transfer learning, where the label space (and hence the task) of the training and testing data are completely disjoint. A similar paradigm is **multitask learning**, where abundant data in the source domain is available and the model attempts to learn both the source and the target tasks. In addition to that, multitask learning also does not assume any scarcity of data in either the source or the target domain. Another variant of the inductive transfer learning is **self-taught learning** [38], when no source domain data is available for training.

3.2 Transductive Transfer Learning

In this kind of transfer learning paradigm, the source and target tasks are not different while the source and target domains can differ. It is assumed that very little or no target domain data is available while training the model. In the context of this paradigm, there can be two possibilities. First, the feature spaces of the source and the target domains may be different, and second, the marginal probability distributions of the source and the target data may differ. The later is a scenario where the source and target domains are different, which is handled by domain adaptation based techniques. Some of the most prominent applications of transductive transfer learning are multi-view object recognition [39, 40], multi-view object detection [41], multi-view facial expression recognition [42], cross-resolution face recognition [15], cross-spectral face recognition [43], and bimodal-vein data mining [44].

3.3   Unsupervised Transfer Learning

When no labeled data in both source and target domains are available, techniques to train an effective model for the target domain falls under the category of unsupervised transfer learning. This refers to unsupervised learning tasks such as clustering [45], dimensionality reduction [46], and so on. Unsupervised transfer learning has been explored in applications such as target detection from hyperspectral images [47], person reidentification [48], object recognition [49], blur and illumination invariant face recognition [50] and classification of breast cancer histopathology images [51].

4   Domain Adaptation

Given the source and target domains, D_s and D_t, source and target and tasks τ_s and τ_t, domain adaptation based techniques aim to learn a model suitable for the target task when $D_s \neq D_t$ and $\tau_s = \tau_t$. In some cases, a few labeled or unlabeled target domain samples X_t may be available. Let us take an example of classification of cars, where we need to classify an image of a car into "k" different categories. A real- world deployment of such an application can be done by the police for monitoring cars. Now, the scenario (daytime/nighttime, amount of traffic, illumination, and speed of traffic) on which the model was trained might need to be diversified with the passage of time. Instead of discarding the old trained model and retraining a new model from scratch, transfer learning techniques aim at adapting the old model with respect to the new data. This ensures that the old knowledge that was learnt is not discarded, rather adapted for the new domain in an efficient manner, which saves time and effort required to train a new model all over again. Thus, domain adaptation is a transductive transfer learning scenario, where the source and target domains are different (Fig. 3). As explained in Sect. 2.1, a domain consists of the data samples/features and marginal distribution of the data. This marginal distribution is a function of the feature space and depends on the feature distribution of the training and testing data. The domain difference of training and testing data can be attributed to the difference in the distribution of data (marginal distribution of the training and testing data) or difference in the feature space of training and testing data. Based on this, domain adaptation may be classified into **Homogeneous** (the former) and **Heterogeneous Domain Adaptation** (the later) types, which are further illustrated as follows.

4.1   Homogeneous Versus Heterogeneous Domain Adaptation

In the Homogeneous Domain Adaptation setting, the feature space of the source and target data are the same, however, the marginal distributions are different, that

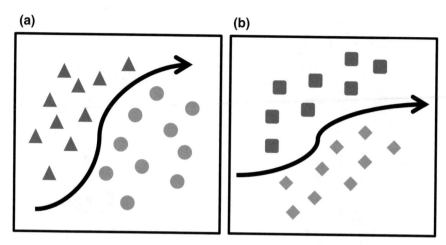

Fig. 3 Illustration of domain adaptation. **a** Represents a classifier (black line) which was trained on the source domain, **b** Represents target domain containing data pertaining to related but different domains, for which the classifier needs to be adapted/fine-tuned

is, $P(X_s) \neq P(X_t)$. The most simple way of performing Homogeneous Domain Adaptation is to fine-tune the pretrained model on labeled instances of the target domain. Several approaches have utilized this technique for low- resolution image classification [54, 55], face verification [56], cross-domain image retrieval [57], face recognition from sketches [58], and cross-spectral face matching [59]. Several methods performed Homogeneous Domain Adaptation by modifying the existing softmax loss to a soft label loss for digit classification [60], object classification [61], and classification of images of cars [62]. Some approaches [56, 63] have also incorporated metric learning into the homogeneous domain adaptation approach. Depending on the availability of labeled target domain data, homogeneous domain adaptation can further be divided into supervised (a few instances of labeled target domain data are available), semi-supervised (abundant unlabeled instances are available), and unsupervised techniques (no labeled data of the target domain is available).

In the heterogeneous setting (Fig. 4), the feature space of the source domain $F(X_s)$ and that of the target domain $F(X_t)$ are different. This scenario may arise due to different models or feature extractors that are used to represent the source and the target domain. One approach to address this is to transform the features of the source and the target domain into a common subspace [64–66]. Another set of approaches has mapped the target domain data into the source domain [67, 68]. Similarly, the heterogeneous scenario may also be classified into supervised, semi-supervised, and unsupervised techniques, depending on the availability of labeled data in the target domain. Homogeneous and heterogeneous domain adaptation are illustrated in Fig. 4.

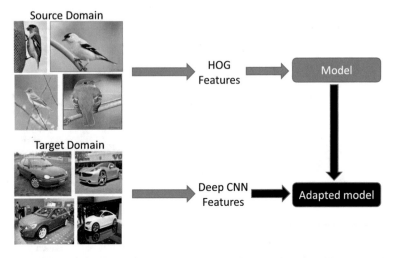

Fig. 4 Illustration of homogeneous and heterogeneous domain adaptation. The source domain consists of images of birds (images taken from CUB-200-2011 database [52]) and the target domain consists of images of cars (images taken from CARS-196 database [53]). As evident in the figure, different feature spaces are used to represent the source domain and target domain, hence the domain adaptation required is heterogeneous. If the same is represented using the identical feature spaces, the domain adaptation required would be homogeneous

4.2 Multistep Domain Adaptation

Domain adaptation techniques in general assume that the source and target domains are related, even if they are not identical. An example of related domains would be face images in high resolution and low resolution while images of cars and faces would be completely unrelated in terms of the marginal distribution of the domains. In cases where the domains are related, the domain adaptation methods are popularly known as **one-step domain adaptation** methods [15, 67–69]. In very limited scenarios, we would have the source and target domains very different from each other. In such cases, **multistep** or transitive domain adaptation approaches [70, 71] needs to be applied.

One-Step Domain Adaptation: One-step domain adaptation (Fig. 5a) is focused on source and target domains that are related. Figure 5 illustrates one-step and multistep domain adaptation using examples of face recognition. One-step domain adaptation methods have been used for problems including low-resolution face recognition [54], face verification [56], digit recognition [60], and object recognition [61, 72]. Some of the methods have utilized it in the semi-supervised setting using pseudo-labels for object recognition [63] and scene recognition [73, 74]. Some attribute-based approaches [61, 62] for one-step domain adaptation have also been proposed for object recognition. Another popular approach for one-step domain adaptation is by

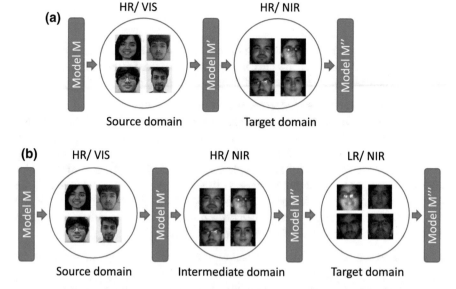

Fig. 5 Illustration of **a** One-step and **b** Multistep domain adaptation. In case of one-step domain adaptation, a model trained on high-resolution (HR) visible spectrum (VIS) face images is adapted to HR and near-infrared (NIR) images since the domains are only different in terms of the spectrum. In case of the multistep approach, the same source domain has to be adapted for LR and NIR images, thus the model is adapted to an intermediate domain (HR/NIR) and then adapted to the final target domain

training the model for the target domain using adversarial training, illustrated as follows.

Adversarial Approaches: Recently, several adversarial approaches have been designed for one-step domain adaptation. In an adversarial approach, samples from the source domain are either transformed into the target domain, or a discriminative model is trained by synthetically generated target domain data. Adversarial-based methods can be based on generative models or non-generative models. Generative model based methods [75–78] use a generative adversarial network (GAN) model for transforming source domain data into target domain data, or generate target domain data from random noise. Non-generative approaches [61, 79–81] learn a model to project source and target domain data into a common feature space so that a domain invariant representation space can be learnt.

Multistep Domain Adaptation: Multistep domain adaptation (Fig. 5a) approaches define one or more intermediate domains for training an effective model for the target domain. As an example, if a model trained on objects is to be adapted for recognition of baby faces, then an intermediate domain could be face images of adults, which can then be adapted to baby faces. Based on the kind of features used, multistep approaches can be classified [82] into handcrafted [83], instance-based [71] or representation transfer based [84] methods.

4.3   Related Areas

Several related areas to domain adaptation such as covariate shift, incremental learning, co-training, and dataset bias have been outlined in transfer learning surveys [34, 85–89]. A concise overview of each of them is as follows:

Covariate Shift or Sample Selection Bias: Covariate shift is a classical problem in transfer learning where the marginal distributions of the two domains are different although the conditional probability of the labels, given the features, are identical. In other words, $P(X_s) \neq P(X_t)$ but $P(Y_s/X = x) == P(Y_t/X = x)$. This setting is known as covariate shift [33] or sample selection bias [90, 91]. One of the major approaches for addressing this scenario is the instance weighting scheme [92]. In order to account for the difference in the marginal distribution of the source and target domain data, weights are assigned to the training samples so that their marginal distribution is similar to that of the target domain data.

Incremental learning: A model trained on a set of data samples may need to be updated for new or upgraded data so the old model's knowledge can be utilized. The new data may be represented by more data samples from the existing classes on which the model was trained, or a new set of classes in addition to the old classes. As an example, a recognition system which identifies car models using images from cameras installed at traffic signals may need to be updated periodically as new car models are out in the market. In order to design an effective incremental learning algorithm [93–95], the plasticity–stability dilemma needs to be addressed. This is a trade-off between catastrophic forgetting [96] (losing the ability to classify the old data/classes) and effectively incorporating the new knowledge in the old model.

Co-training: In order to train a model for a real-world application, machine learning algorithms use labeled data for training. In such cases, scarcity of labeled data may be a hindrance to training an effective model. Most often, abundant unlabeled data is available, which can be utilized for training the model. Co-training algorithms [28, 97, 98] use these unlabeled samples in conjunction with the labeled samples to update/learn the classifiers. The approach is to have two classifiers/models trained on separate views of data. This helps in the representation of complementary knowledge which can be utilized to assign pseudo-labels to the unlabeled samples to aid training of the classifiers.

5   Summary

This chapter summarizes the different aspects of domain adaptation for visual understanding along with related disciplines and applications. Domain adaptation based techniques present an important set of tools which are utilized to train effective machine learning models for which abundant target domain data is not available. It falls under the transductive transfer learning paradigm, where the source and target tasks are the same but the source and target domains are different. With the recent usage and popularity of deep representation learning methods and the advent of large-scale data based applications, the relevance and importance of domain adaptation of pretrained models is high. A significant amount of research has been done in Homogeneous and one-step domain adaptation. However, the potential of Heterogeneous and multistep domain adaptation approaches for visual understanding is relatively unexplored.

References

1. Zhao W, Chellappa R, Phillips PJ, Rosenfeld A (2003) Face recognition: a literature survey. ACM Comput Surv 35(4):399–458
2. Singh R, Vatsa M, Ross A, Noore A (2010) Biometric classifier update using online learning: a case study in near infrared face verification. Image Vis Comput 28(7):1098–1105
3. Bharadwaj S, Bhatt HS, Singh R, Vatsa M, Noore A (2015) Qfuse: online learning framework for adaptive biometric system. Pattern Recognit 48(11):3428–3439
4. Singh R, Vatsa M, Ross A, Noore A (2009) Online learning in biometrics: a case study in face classifier update. In: International conference on biometrics: theory, applications, and systems, pp 1–6
5. Mehrotra H, Singh R, Vatsa M, Majhi B (2016) Incremental granular relevance vector machine: a case study in multimodal biometrics. Pattern Recognit 56:63–76
6. Chen JC, Patel VM, Chellappa R (2016) Unconstrained face verification using deep CNN features. In: IEEE winter conference on applications of computer vision, pp 1–9
7. Chen YC, Patel VM, Phillips PJ, Chellappa R (2012) Dictionary-based face recognition from video. In: European conference on computer vision, pp 766–779
8. Felzenszwalb PF, Girshick RB, McAllester D, Ramanan D (2010) Object detection with discriminatively trained part-based models. IEEE Trans Pattern Anal Mach Intell 32(9):1627–1645
9. Yilmaz A, Javed O, Shah M (2006) Object tracking: a survey. ACM Comput Surv 38(4):13
10. Li LJ, Socher R, Fei-Fei L (2009) Towards total scene understanding: classification, annotation and segmentation in an automatic framework. In: IEEE conference on computer vision and pattern recognition, pp 2036–2043
11. Rautaray SS, Agrawal A (2015) Vision based hand gesture recognition for human computer interaction: a survey. Artif Intell Rev 43(1):1–54
12. da Fontoura Costa L, Cesar RM Jr (2010) Shape analysis and classification: theory and practice. CRC Press, Boca Raton
13. Rui Y, Huang TS, Chang SF (1999) Image retrieval: current techniques, promising directions, and open issues. J Vis Commun Image Represent 10(1):39–62
14. Bao L, Intille SS (2004) Activity recognition from user-annotated acceleration data. In: International conference on pervasive computing, pp 1–17
15. Bhatt HS, Singh R, Vatsa M, Ratha NK (2014) Improving cross-resolution face matching using ensemble-based co-transfer learning. IEEE Trans Image Process 23(12):5654–5669
16. Singh M, Nagpal S, Vatsa M, Singh R, Majumdar A (2018) Identity aware synthesis for cross resolution face recognition. In: IEEE conference on computer vision and pattern recognition workshops, pp 479–488
17. Dhamecha TI, Sharma P, Singh R, Vatsa M (2014) On effectiveness of histogram of oriented gradient features for visible to near infrared face matching. In: International conference on pattern recognition, pp 1788–1793
18. Ghosh S, Dhamecha TI, Keshari R, Singh R, Vatsa M (2015) Feature and keypoint selection for visible to near-infrared face matching. In: International conference on biometrics theory, applications and systems, pp 1–7

19. Mudunuri SP, Biswas S (2016) Low resolution face recognition across variations in pose and illumination. IEEE Trans Pattern Anal Mach Intell 38(5):1034–1040
20. Yadav D, Singh R, Vatsa M, Noore A (2014) Recognizing age-separated face images: humans and machines. PloS One 9(12):1122–1134
21. Dhamecha TI, Singh R, Vatsa M, Kumar A (2014) Recognizing disguised faces: human and machine evaluation. PloS One 9(7):e99212
22. Kushwaha V, Singh M, Singh R, Vatsa M, Ratha N, Chellappa R (2018) Disguised faces in the wild. In: IEEE conference on computer vision and pattern recognition workshops, pp 1–9
23. Nguyen HV, Ho HT, Patel VM, Chellappa R (2015) Dash-n: joint hierarchical domain adaptation and feature learning. IEEE Trans Image Process 24(12):5479–5491
24. Shrivastava A, Shekhar S, Patel VM (2014) Unsupervised domain adaptation using parallel transport on grassmann manifold. In: IEEE winter conference on applications of computer vision, pp 277–284
25. Shekhar S, Patel VM, Nguyen HV, Chellappa R (2013) Generalized domain-adaptive dictionaries. In: 2013 IEEE conference on computer vision and pattern recognition, pp 361–368
26. Qiu Q, Patel VM, Turaga P, Chellappa R (2012) Domain adaptive dictionary learning. In: European conference on computer vision, pp 631–645
27. Zhang H, Patel VM, Shekhar S, Chellappa R (2015) Domain adaptive sparse representation-based classification. In: 2015 11th IEEE international conference and workshops on automatic face and gesture recognition (FG), vol 1, pp 1–8
28. Bharadwaj S, Bhatt HS, Vatsa M, Singh R (2016) Domain specific learning for newborn face recognition. IEEE Trans Inf Forensics Secur 11(7):1630–1641
29. Bharadwaj S, Bhatt HS, Singh R, Vatsa M, Singh SK (2010) Face recognition for newborns: a preliminary study. In: IEEE international conference on biometrics: theory, applications and systems, pp 1–6
30. Yin X, Han J, Yang J, Philip SY (2006) Efficient classification across multiple database relations: a crossmine approach. IEEE Trans Knowl Data Eng 18(6):770–783
31. Kuncheva LI, Rodriguez JJ (2007) Classifier ensembles with a random linear oracle. IEEE Trans Knowl Data Eng 19(4):500–508
32. Baralis E, Chiusano S, Garza P (2008) A lazy approach to associative classification. IEEE Trans Knowl Data Eng 20(2):156–171
33. Shimodaira H (2000) Improving predictive inference under covariate shift by weighting the log-likelihood function. J Stat Plan Inference 90(2):227–244
34. Pan SJ, Yang Q et al (2010) A survey on transfer learning. IEEE Trans Knowl Data Eng 22(10):1345–1359
35. Socher R, Ganjoo M, Manning CD, Ng A (2013) Zero-shot learning through cross-modal transfer. In: Advances in neural information processing systems, pp 935–943
36. Palatucci M, Pomerleau D, Hinton GE, Mitchell TM (2009) Zero-shot learning with semantic output codes. In: Advances in neural information processing systems, pp 1410–1418
37. Romera-Paredes B, Torr P (2015) An embarrassingly simple approach to zero-shot learning. In: International conference on machine learning, pp 2152–2161
38. Raina R, Battle A, Lee H, Packer B, Ng AY (2007) Self-taught learning: transfer learning from unlabeled data. In: International conference on machine learning, pp 759–766
39. Blanz V, Schölkopf B, Bülthoff H, Burges C, Vapnik V, Vetter T (1996) Comparison of view-based object recognition algorithms using realistic 3D models. In: International joint conference on artificial intelligence, pp 251–256
40. LeCun Y, Huang FJ, Bottou L (2004) Learning methods for generic object recognition with invariance to pose and lighting. In: IEEE conference on computer vision and pattern recognition, vol 2, pp 97–104
41. Liebelt J, Schmid C (2010) Multi-view object class detection with a 3D geometric model. In: IEEE conference on computer vision and pattern recognition, pp 1688–1695
42. Moore S, Bowden R (2011) Local binary patterns for multi-view facial expression recognition. Elsevier Comput Vis Image Underst 115(4):541–558

43. Juefei-Xu F, Pal DK, Savvides M (2015) NIR-VIS heterogeneous face recognition via cross-spectral joint dictionary learning and reconstruction. In: IEEE conference on computer vision and pattern recognition, pp 141–150
44. Wang J, Wang G, Zhou M (2018) Bimodal vein data mining via cross-selected-domain knowledge transfer. IEEE Trans Inf Forensics Secur 13(3):733–744
45. Dai W, Yang Q, Xue GR, Yu Y (2008) Self-taught clustering. In: International conference on machine learning, pp 200–207
46. Wang Z, Song Y, Zhang C (2008) Transferred dimensionality reduction. In: Joint European conference on machine learning and knowledge discovery in databases, pp 550–565
47. Du B, Zhang L, Tao D, Zhang D (2013) Unsupervised transfer learning for target detection from hyperspectral images. Elsevier Neurocomputing 120:72–82
48. Peng P, Xiang T, Wang Y, Pontil M, Gong S, Huang T, Tian Y (2016) Unsupervised cross-dataset transfer learning for person re-identification. In: IEEE conference on computer vision and pattern recognition, pp 1306–1315
49. Pinheiro PO, Element A (2017) Unsupervised domain adaptation with similarity learning. In: IEEE conference on computer vision and pattern recognition, pp 8004–8013
50. Yang B, Ma AJ, Yuen PC (2018) Learning domain-shared group-sparse representation for unsupervised domain adaptation. Elsevier Pattern Recognit 81:615–632
51. Alirezazadeh P, Hejrati B, Monsef-Esfehani A, Fathi A (2018) Representation learning-based unsupervised domain adaptation for classification of breast cancer histopathology images. Elsevier Biocyber Biomed Eng 38(3):671–683
52. Wah C, Branson S, Welinder P, Perona P, Belongie S (2011) The Caltech-UCSD birds-200-2011 dataset. Technical Report CNS-TR-2011-001, California Institute of Technology
53. Krause J, Stark M, Deng J, Fei-Fei L (2013) 3D object representations for fine-grained categorization. In: IEEE workshop on 3D representation and recognition (3dRR-13), Sydney, Australia
54. Peng X, Hoffman J, Yu SX, Saenko K (2016) Fine-to-coarse knowledge transfer for low-res image classification. arXiv:1605.06695
55. Yao Y, Li X, Ye Y, Liu F, Ng MK, Huang Z, Zhang Y (2018) Low-resolution image categorization via heterogeneous domain adaptation. Knowl Based Syst 163:656–665
56. Hu J, Lu J, Tan YP (2015) Deep transfer metric learning. In: IEEE conference on computer vision and pattern recognition, pp 325–333
57. Wang X, Duan X, Bai X (2016) Deep sketch feature for cross-domain image retrieval. Elsevier Neurocomputing 207:387–397
58. Mittal P, Vatsa M, Singh R (2015) Composite sketch recognition via deep network-a transfer learning approach. In: IAPR international conference on biometrics, pp 251–256
59. Liu X, Song L, Wu X, Tan T (2016) Transferring deep representation for NIR-VIS heterogeneous face recognition. In: IAPR international conference on biometrics, pp 1–8
60. Hinton G, Vinyals O, Dean J (2015) Distilling the knowledge in a neural network. arXiv:1503.02531
61. Tzeng E, Hoffman J, Darrell T, Saenko K (2015) Simultaneous deep transfer across domains and tasks. In: IEEE international conference on computer vision, pp 4068–4076
62. Gebru T, Hoffman J, Fei-Fei L (2017) Fine-grained recognition in the wild: a multi-task domain adaptation approach. In: International conference on computer vision, pp 1358–1367
63. Motiian S, Piccirilli M, Adjeroh DA, Doretto G (2017) Unified deep supervised domain adaptation and generalization. In: IEEE international conference on computer vision, pp 5715–5725
64. Duan L, Xu D, Tsang I (2012) Learning with augmented features for heterogeneous domain adaptation. arXiv:1206.4660
65. Wang C, Mahadevan S (2011) Heterogeneous domain adaptation using manifold alignment. In: International joint conference on artificial intelligence, vol 22, pp 1541–1546
66. Zhou JT, Tsang IW, Pan SJ, Tan M (2014) Heterogeneous domain adaptation for multiple classes. In: Artificial intelligence and statistics, pp 1095–1103
67. Kulis B, Saenko K, Darrell T (2011) What you saw is not what you get: domain adaptation using asymmetric kernel transforms. In: IEEE conference on computer vision and pattern recognition, pp 1785–1792

68. Saenko K, Kulis B, Fritz M, Darrell T (2010) Adapting visual category models to new domains. In: European conference on computer vision, pp 213–226
69. Jhuo IH, Liu D, Lee D, Chang SF (2012) Robust visual domain adaptation with low-rank reconstruction. In: IEEE conference on computer vision and pattern recognition, pp 2168–2175
70. Tan B, Song Y, Zhong E, Yang Q (2015) Transitive transfer learning. In: International conference on knowledge discovery and data mining, pp 1155–1164
71. Tan B, Zhang Y, Pan SJ, Yang Q (2017) Distant domain transfer learning. In: AAAI conference on artificial intelligence, pp 2604–2610
72. Baktashmotlagh M, Harandi MT, Lovell BC, Salzmann M (2013) Unsupervised domain adaptation by domain invariant projection. In: IEEE international conference on computer vision, pp 769–776
73. Long M, Zhu H, Wang J, Jordan MI (2016) Unsupervised domain adaptation with residual transfer networks. In: Advances in neural information processing systems, pp 136–144
74. Zhang X, Yu FX, Chang SF, Wang S (2015) Deep transfer network: unsupervised domain adaptation. arXiv:1503.00591
75. Bousmalis K, Silberman N, Dohan D, Erhan D, Krishnan D (2017) Unsupervised pixel-level domain adaptation with generative adversarial networks. In: IEEE conference on computer vision and pattern recognition, pp 3722–3731
76. Liu MY, Tuzel O (2016) Coupled generative adversarial networks. In: Advances in neural information processing systems, pp 469–477
77. Isola P, Zhu JY, Zhou T, Efros AA (2017) Image-to-image translation with conditional adversarial networks. In: International conference on computer vision, pp 1125–1134
78. Yi Z, Zhang HR, Tan P, Gong M (2017) Dualgan: unsupervised dual learning for image-to-image translation. In: International conference on computer vision, pp 2868–2876
79. Tzeng E, Devin C, Hoffman J, Finn C, Abbeel P, Levine S, Saenko K, Darrell T (2015) Adapting deep visuomotor representations with weak pairwise constraints. arXiv:1511.07111
80. Ganin Y, Lempitsky V (2014) Unsupervised domain adaptation by backpropagation. arXiv:1409.7495
81. Ganin Y, Ustinova E, Ajakan H, Germain P, Larochelle H, Laviolette F, Marchand M, Lempitsky V (2016) Domain-adversarial training of neural networks. J Mach Learn Res 17(1):2096–2130
82. Wang M, Deng W (2018) Deep visual domain adaptation: a survey. Neurocomputing 312:135–153
83. Xie M, Jean N, Burke M, Lobell D, Ermon S (2015) Transfer learning from deep features for remote sensing and poverty mapping. arXiv:1510.00098
84. Rusu AA, Rabinowitz NC, Desjardins G, Soyer H, Kirkpatrick J, Kavukcuoglu K, Pascanu R, Hadsell R (2016) Progressive neural networks. arXiv:1606.04671
85. Csurka G (2017) Domain adaptation for visual applications: a comprehensive survey. arXiv:1702.05374
86. Patel VM, Gopalan R, Li R, Chellappa R (2015) Visual domain adaptation: a survey of recent advances. IEEE Signal Process Mag 32(3):53–69
87. Shao L, Zhu F, Li X (2015) Transfer learning for visual categorization: a survey. IEEE Trans Neural Netw Learn Syst 26(5):1019–1034
88. Zhang J, Li W, Ogunbona P (2017) Transfer learning for cross-dataset recognition: a survey. arXiv:1705.04396
89. Zhang L (2019) Transfer adaptation learning: a decade survey. arXiv:1903.04687
90. Heckman J et al (2013) Sample selection bias as a specification error. Appl Econ 31(3):129–137
91. Zadrozny B (2004) Learning and evaluating classifiers under sample selection bias. In: International conference on machine learning, p 114–122
92. Jiang J (2008) Domain adaptation in natural language processing. Technical report
93. Zhao H, Yuen PC (2008) Incremental linear discriminant analysis for face recognition. IEEE Trans Syst Man Cybern Part B (Cybernetics) 38(1):210–221
94. Liu LP, Jiang Y, Zhou ZH (2009) Least square incremental linear discriminant analysis. In: IEEE international conference on data mining, pp 298–306

95. Xiao T, Zhang J, Yang K, Peng Y, Zhang Z (2014) Error-driven incremental learning in deep convolutional neural network for large-scale image classification. In: ACM international conference on multimedia, pp 177–186
96. Kirkpatrick J, Pascanu R, Rabinowitz N, Veness J, Desjardins G, Rusu AA, Milan K, Quan J, Ramalho T, Grabska-Barwinska A et al (2017) Overcoming catastrophic forgetting in neural networks. In: Proceedings of the national academy of sciences, vol 114(13), pp 3521–3526
97. Bhatt HS, Bharadwaj S, Singh R, Vatsa M, Noore A, Ross A (2011) On co-training online biometric classifiers. In: International joint conference on biometrics, pp 1–7
98. Blum A, Mitchell T (1998) Combining labeled and unlabeled data with co-training. In: Annual conference on computational learning theory, pp 92–100

M-ADDA: Unsupervised Domain Adaptation with Deep Metric Learning

Issam H. Laradji and Reza Babanezhad

Abstract Unsupervised domain adaptation techniques have been successful for a wide range of problems where supervised labels are limited. The task is to classify an unlabeled "target" dataset by leveraging a labeled "source" dataset that comes from a slightly similar distribution. We propose metric-based adversarial discriminative domain adaptation (M-ADDA) which performs two main steps. First, it uses a metric learning approach to train the source model on the source dataset by optimizing the triplet loss function. This results in clusters where embeddings of the same label are close to each other and those with different labels are far from one another. Next, it uses the adversarial approach (as that used in ADDA (Tzeng et al. Adversarial discriminative domain adaptation, 2017, [36])) to make the extracted features from the source and target datasets indistinguishable. Simultaneously, we optimize a novel loss function that encourages the target dataset's embeddings to form clusters. While ADDA and M-ADDA use similar architectures, we show that M-ADDA performs significantly better on the digits adaptation datasets of MNIST and USPS. This suggests that using metric learning for domain adaptation can lead to large improvements in classification accuracy for the domain adaptation task. The code is available at https://github.com/IssamLaradji/M-ADDA.

Keywords Domain adaptation · Metric learning · Triplet loss · Adversarial learning

I. H. Laradji (✉) · R. Babanezhad
Department of Computer Science, University of British Columbia,
Vancouver, BC, Canada
e-mail: issamou@cs.ubc.ca

R. Babanezhad
e-mail: rezababa@cs.ubc.ca

© Springer Nature Switzerland AG 2020
R. Singh et al. (eds.), *Domain Adaptation for Visual Understanding*,
https://doi.org/10.1007/978-3-030-30671-7_2

17

1 Introduction

Convolutional neural networks (CNN) [19] allow us to extract powerful features that can be used for tasks such as image classification and segmentation. However, these features are usually domain specific in that they are not discriminative enough for datasets coming from other domains, resulting in poor classification performance. Consequently, unsupervised domain adaptation techniques have emerged [9, 21, 35, 38] to address the domain shift phenomenon between a source dataset and a target dataset. Common techniques use adversarial learning in order to make extracted features from the source and target datasets indistinguishable. The extracted features from the target dataset are then passed through a trained classifier (pretrained on the source dataset) to predict the labels of the target test set [36].

Recently, metric-based methods have been introduced to address the problem of unsupervised domain adaptation [14, 24], namely classifying an example is performed by computing its similarity to prototype representations of each category [24]. Further, a category-agnostic clustering network was proposed by [14] to cluster new datasets through transfer learning. In this chapter, we introduce M-ADDA, a metric-based adversarial discriminative domain adaptation framework. First, M-ADDA trains our source model using metric learning by optimizing the triplet loss [13] on the source dataset. As a result, if K is the number of classes, then the dataset is clustered into K clusters where each cluster is composed of examples having the same label (see Fig. 1). The goal is to obtain an embedding of the target dataset where the k-nearest neighbors (kNN) of each example belongs to the same class and where examples from different classes are separated by a large margin. A major strength

Fig. 1 Metric learning. The result of minimizing the triplet loss on the MNIST dataset. Each cluster corresponds to examples belonging to a single-digit label

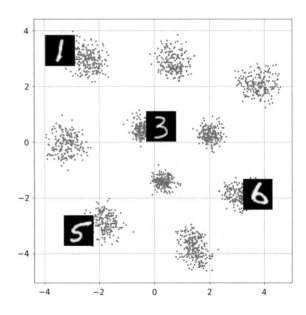

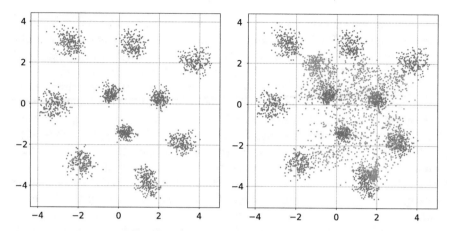

Fig. 2 Domain adaptation. The **blue** dots represent the MNIST embeddings after optimizing Eq. (1). The **orange** dots represent the USPS embeddings. The center image shows the USPS embeddings before minimizing the domain shift adverbially by Eq. (3). The right-most image shows the USPS embeddings after optimizing Eq. (2)

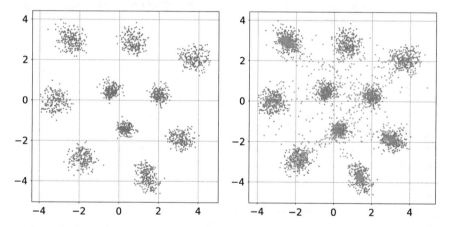

Fig. 3 Domain adaptation. The **blue** dots represent the MNIST embeddings after optimizing Eq. (1). The **orange** dots represent the USPS embeddings. The center image shows the USPS embeddings before minimizing the domain shift adverbially by Eq. (3). The right-most image shows the USPS embeddings after optimizing Eq. (2)

in this approach is its nonparametric nature [39] as it does not implicitly make parametric (possibly limiting) assumptions about the input distributions (Figs. 2 and 3).

Next, we adapt the distributions between the source and target extracted features using the adversarial learning method used by ADDA [36]. This addresses the domain discrepancy between the datasets. Early methods for domain adaptation are based on minimizing correlation distances and minimizing the maximum mean discrepancy to ensure both datasets have a common feature space [22, 31, 32, 34].

However, adversarial learning approaches showed state-of-the-art performance for domain adaptation. While the features' distributions become more similar during training, we also train a network that maps the extracted features to embeddings such that they are clustered into K clusters. Concurrently, we encourage the clusters to have large margins between them. Therefore, the network is trained by minimizing the distance between each target example embedding and its closest cluster center corresponding to the source embedding. This approach is simple to implement and achieves competitive results on digit datasets such as MNIST [18] and USPS [17].

To summarize our contributions, (1) we propose a novel metric learning framework that uses the triplet loss to cluster the source dataset for the task of domain adaptation; (2) we propose a new loss function that regularizes the embeddings of the target dataset to encourage them to form clusters; and (3) we show a large improvement over ADDA [36] on a standard unsupervised domain adaptation benchmark. Note that ADDA uses a similar architecture but a different loss function than M-ADDA.

In Sect. 2, we review the related works and other similar approaches. In Sect. 3, we introduce our framework and the new loss terms for domain adaptation. In Sect. 4, we present experimental results illustrating the efficacy of our approach on the digits dataset. Finally, we conclude the chapter in Sect. 5.

2 Related Work

Metric learning has shown great success in many visual classification tasks [13, 30, 39]. The goal is to learn a distance metric such that examples belonging to the same label are as close as possible in some embedding space and samples from different labels are as far from one another as possible. It can be used for unsupervised learning such as clustering [40] and supervised learning such as k-nearest neighbor algorithms [12, 39]. Recently, triplet networks [13] and Siamese networks [1] have been proposed as powerful models for metric learning which have been successfully applied for few-shot learning and learning with few data. However, to the best of our knowledge, we are the first to apply metric learning that is based on triplet networks for domain adaptation.

A close topic to domain adaptation is *transfer learning* which has received tremendous attention recently. It allows us to solve tasks where labels are scarce by learning from relevant tasks for which labels are abundant [4, 5, 28] by identifying a common structure between multiple tasks [6]. A common transfer learning strategy is to use pretrained networks such as those trained on ImageNet [15] and fine-tune them on new tasks. While this approach can significantly improve performance for many visual tasks, it performs poorly when the pretrained network is used on a dataset which comes from a different distribution than the one it is trained on. This is because the model has learned features that are specific to one domain that might not be meaningful for other domains.

To address this challenge, a large set of domain adaptation methods were proposed over the years [9, 21, 35, 36] whose goal is to determine a common latent space between two domains often referred to as a source dataset and a target dataset. The general setting is to use a model that trains to extract features from the source dataset, and then encourage features extracted from the target dataset to be similar to the source features [2, 8, 10, 26, 37]. Auto-encoder based methods [3, 10] train one or a variety of auto-encoders for the source and target datasets. Then, a classifier is trained based on the latent representation of the source dataset. The same classifier is then used to label the target dataset. Adversarial networks [11] based approaches use a generator model to transform the examples' feature representations from one domain to another [3, 25, 29].

Another group of domain adaptation methods [20, 31, 34, 37] minimize the difference between the distributions of the features extracted from the source and target data. They achieve this by minimizing point estimates of a given metric between the source and target distributions by using maximum or mean discrepancy metrics. Current state-of-the-art techniques use the adversarial learning approach to encourage the feature representations from the two datasets to be indistinguishable (i.e., have a common distribution) [36]. Close to our method are recent similarity-based approaches proposed by [14, 24], which transfer class-agnostic prior to new datasets, and classify examples by computing their similarity to prototype representation of each category, respectively. Our approach uses a regularized metric learning method with the help of k-nearest neighbors as a nonparametric framework. This can be more powerful than ADDA which uses a model that makes parametric assumptions (introducing limitations) about the input distribution [39].

Another class of domain adaptation methods are self-ensembling methods which augment the source dataset by applying various label preserving transformations on the images [7, 16, 27, 33]. Using the augmented dataset, they train several deep network models and use an ensemble of those networks for the domain adaptation task. Laine et al. [16] have two networks in their model: the Π-model and temporal model. In the Pi-model, every unlabelled sample feeds to a classifier twice with different dropout, noise, and image translation parameters. Their temporal model records the average of the historical network prediction per sample and forces the subsequent predictions to be close to the average. Tarvainen et al. [33] improve the temporal network by recording the average of the network weights rather than class prediction. This results in two networks: the student and the teacher network. The student network is trained via gradient descent and the weights of the teacher are the historical exponential moving average of the weights of the student network. The unsupervised loss is the mean square difference between the prediction of the student and the teacher under different dropout, noise, and image translation parameters. French et al. [7] combine the previous two methods with adding extra modifications and engineering and get state-of-the-art results in many domain adaptation tasks for image datasets. However, this method uses heavy engineering with many label preserving transformations to augment the data. In contrast, we show that our method significantly improves results over ADDA by making simple changes to their framework.

3   Proposed Approach: M-ADDA

We propose M-ADDA which performs two main steps:

1. train a source model on the source dataset using metric learning (as in Fig. 4) using the triplet loss function; then
2. simultaneously, adapt the distributions between the extracted source and target dataset features and regularize the predicted target dataset embeddings to form clusters (see Fig. 5).

Our M-ADDA framework consists of a source model and a target model. The two models have the same architecture, and they both have an encoder that extracts features from the input dataset and a decoder to map the extracted features to embeddings. Consider a source dataset (X_S, Y_S), and a target dataset (X_T, Y_T) where the data X_S and X_T are drawn from two different distributions.

Training the source model. The source model $f_{\theta_S}(\cdot)$, parameterized by θ_S, is first trained on the source dataset by optimizing the following triplet loss:

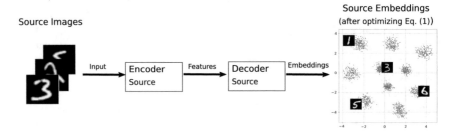

Fig. 4   Training the source model. We pretrain the source encoder and decoder by optimizing the triplet loss in Eq. (1). The source encoder extracts the features from the source dataset and the decoder maps the features to the embedding space where clusters are formed

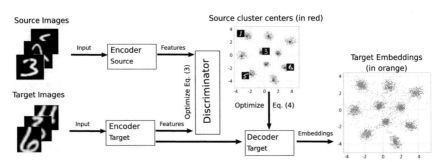

Fig. 5   Training the target model. We adversarially adapt the encoded features' distributions between the source and target encoder using Eq. (3) while using the source cluster centers to optimize Eq. (4). The label of each target embedding is the mode of the labels of the nearest source embedding neighbors

$$\mathcal{L}(\theta_S) = \sum_{(a_i, p_i, n_i)} \max(||f_{\theta_S}(a_i) - f_{\theta_S}(p_i)||^2 - \tag{1}$$

$$||f_{\theta_S}(a_i) - f_{\theta_S}(n_i)||^2 + m, 0)$$

where a_i is an anchor example (picked randomly), p_i is an example with the same label as the anchor, and n_i is an example with a different label from the anchor. Optimizing Eq. (1) encourages the embedding of a_i to be closer to p_i than to n_i by at least margin m. If the anchor example is close enough to the positive example p_i, and far from the negative example n_i by a margin of at least m, the *max* function returns zero; therefore, the corresponding triplet (a_i, p_i, n_i) does not contribute to the loss function. If the margin is smaller than m, then the *max* function returns $||f_{\theta_S}(a_i) - f_{\theta_S}(p_i)||^2 - ||f_{\theta_S}(a_i) - f_{\theta_S}(n_i)||^2 + m$. Minimizing this term results in moving a_i toward p_i and moving it away from n_i in the embedding feature space. After optimizing the loss term long enough, the samples with the same label are pulled together and those with different labels are pushed away from each other. As a result, points of the same label form a single cluster which allows us to efficiently classify examples using k-nearest neighbors (see Fig. 4).

Algorithm 1 shows the procedure of training the source model on the source dataset for one epoch. Given a batch (X_B, Y_B), for each unique element y_i in Y_B, we obtain an anchor a_i whose label is y_i, a positive example p_i whose label is y_i, and a negative example n_i whose label is not y_i. Note that set(Y_B) returns the unique elements of Y_B. In our experiments, we obtained the negative example uniformly at random. However, other methods are possible such as greedily picking the triplet with the largest loss (as computed by Eq. (1)), and non-uniformly picking triplets based on their individual loss values. Finally, for each triplet, we compute the loss and update the parameters of the source model to minimize Eq. (1).

Algorithm 1 Training the source model on the source dataset (single epoch).

1: **inputs**
2:    Source model $f_{\theta_S}(\cdot)$, and source images and labels (X_S, Y_S).
3: **for** $\{X_B, Y_B\} \in (X_S, Y_S)$ **do**
4:    **for** $y_i \in$ set(Y_B) **do**
5:        $AP \leftarrow$ All image pairs whose label is y_i.
6:        **for** each $\{a_i, p_i\} \in AP$ **do**
7:            $n_i \leftarrow$ A random sample in X_B whose label is not y_i.
8:            $L \leftarrow$ The loss in Eq. (1) using $\{a_i, p_i, n_i\}$ and $f_{\theta_S}(\cdot)$.
9:            Update the parameters θ_S by backpropagating through L.
10:       **end for**
11:   **end for**
12: **end for**

Training the target model. Next, we define C as the set of centers corresponding to the source embedding clusters (represented as red dots in Fig. 5). Each center in C corresponds to a single label in the source dataset. A center is computed by taking the mean of the source embeddings belonging to that center's label. Then, we train the target model, parametrized by θ_T by optimizing the following two loss terms:

$$\mathcal{L}(\theta_T, \theta_D) = \underbrace{\mathcal{L}_A(\theta_{T_E}, \theta_D)}_{\text{Adapt}} + \underbrace{\mathcal{L}_C(\theta_T)}_{\text{C-Magnet}} \tag{2}$$

where θ_{T_E} corresponds to the parameters of the target model's encoder; and θ_D is the parameter set for a discriminator model we use to adapt the distributions of the extracted features between the source (S) and target (T) datasets. We achieve this by optimizing

$$\mathcal{L}_A(\theta_{T_E}, \theta_D) = \min_{\theta_D} \max_{\theta_{T_E}} - \sum_{i \in S} \log D_{\theta_D}(E_{\theta_S}(X_{S_i})) - \\ \sum_{i \in T} \log (1 - D_{\theta_D}(E_{\theta_{T_E}}(X_{T_i}))), \tag{3}$$

where θ_{S_E} is the source model encoder's set of parameters; and $D(\cdot)$ is the discriminator model which is trained to maximize the probability that the features extracted by the source model's encoder come from the source dataset and that the features extracted by the target model's encoder come from the target dataset. In other words, the discriminator $D(.)$ tries to distinguish between the features extracted from the source dataset and the features from the target dataset by giving higher value (close to one) to a source dataset feature vector and a lower value (close to zero) to a target dataset feature vector. Simultaneously, the encoder of the target model is trained to confuse the discriminator into predicting the target features as coming from the source dataset. This adversarial learning approach encourages the features extracted by $E_{\theta_{S_E}}(X_{S_i})$ and $E_{\theta_{T_E}}(X_{T_i})$ to be indistinguishable in their distributions. For the sake of brevity, note that we show the loss functions in terms of a single source example X_{S_i} and target example X_{T_i}.

In parallel, we minimize the center magnet loss term defined as

$$\mathcal{L}_C(\theta_T) = \sum_{i \in T} \min_{j} ||f_{\theta_T}(x_i) - C_j||^2, \tag{4}$$

which pulls the embeddings of example X_i to the closest cluster center defined in C (see Fig. 5). The cluster center for a class is obtained by taking the Euclidean mean of all samples belonging to that class. Since we have 10 classes in MNIST and USPS, $|C| = 10$. This regularization term allows the target dataset embeddings to form clusters that are similar to the clusters formed by the source dataset embeddings. This is useful when minimizing $\mathcal{L}(\theta_T, \theta_D)$ fails to make the target embedding clustered in a similar way as the source embeddings. For example, in Fig. 2b, we see that the target embeddings become scattered around the center when minimizing $\mathcal{L}_A(\theta_T, \theta_D)$ only. However, by simultaneously minimizing $\mathcal{L}_C(\theta_T)$, we get a better formation of clusters as seen in Fig. 3b.

Algorithm 2 Training the target model on the target dataset (single epoch).

1: **inputs**
2:   Target model $f_{\theta_T}(\cdot)$, and source and target images and labels (X_S, Y_S, X_T, Y_T).
3: **for** $\{X_{S_B}, Y_{S_B}, X_{T_B}, Y_{T_B}\} \in (X_S, Y_S, X_T, Y_T)$ **do**
4:   Maximize Eq. (3) w.r.t. θ_D using $\{X_{S_B}, Y_{S_B}, X_{T_B}, Y_{T_B}\}$
5:   Minimize Eq. (3) w.r.t. θ_T using $\{X_{S_B}, Y_{S_B}, X_{T_B}, Y_{T_B}\}$
6: **end for**
7: $E_S \leftarrow$ The embeddings of the source dataset extracted by $f_{\theta_S}(\cdot)$
8: $C \leftarrow$ The cluster centers of E_S are obtained by taking the Euclidean mean for each class.
9: **for** $\{X_{T_B}, Y_{T_B}\} \in (X_T, Y_T)$ **do**
10:   $L \leftarrow$ The loss computed using Eq. (4) and cluster centers C
11:   Update parameters θ_T by backpropagating through L.
12: **end for**

Algorithm 3 Predicting the labels of the test images.

1: **inputs**
2:   Target model $f_{\theta_T}(\cdot)$, Source model $f_{\theta_T}(\cdot)$, and source and target images and labels.
3: $E_S \leftarrow$ The embeddings of the source dataset extracted by $f_{\theta_S}(\cdot)$
4: **for** $\{X_{T_B}, Y_{T_B}\} \in (X_T, Y_T)$ **do**
5:   $E_{T_B} \leftarrow$ The embeddings of X_{T_B} extracted by $f_{\theta_T}(\cdot)$
6:   $P_{T_B} \leftarrow$ The mode label of the k-nearest E_S samples.
7: **end for**

Algorithm 2 shows the procedure for training the target model on the target dataset. Lines 4–5 use Eq. (3) to make the target features and the source features indistinguishable. Lines 7–12 update the target model parameters by encouraging the target embeddings to move to the closest source cluster center. As shown in Algorithm 3, the prediction stage consists of two steps. First, we extract the embeddings of the source dataset examples using the pretrained source model. Then, the label of an example X_{T_i} is the mode label of the k-nearest source embeddings. This nonparametric approach allows us to implicitly learn powerful features that are used to compute the similarities between the examples.

4   Experiments

To illustrate the performance of our method for the unsupervised domain adaptation task, we apply it on the standard digits dataset benchmark using accuracy as the evaluation metric. We consider two domains: MNIST and USPS. They consist of 10 classes representing the digits between 0 and 9 (we show some digit examples in Fig. 6). We follow the experimental setup in [36] where 2000 images are sampled from MNIST and 1800 from USPS for training. Since our task is unsupervised domain adaptation, all the images in the target domain are unlabeled. In each experiment, we ran Algorithm 1 for 200 epochs to train our source model. Then, we reported the accuracy on the target test set after running Algorithm 2 for 200 epochs.

MNIST
USPS

Fig. 6 Dataset. Example images taken from the two digit domains we used in our benchmark

Table 1 Digits adaptation. We evaluate our method on the unsupervised domain adaptation task on the digits datasets, using the setup in [36]

Method	MNIST → USPS	USPS → MNIST
Source only (ADDA [36])	0.752	0.571
Source only (Ours)	0.601	0.679
Gradient reversal [9]	0.771	0.730
Domain confusion [35]	0.791	0.665
CoGAN [21]	0.912	0.891
ADDA [36]	0.894	0.901
M-ADDA (Ours)	**0.952**	**0.940**

Table 2 Digits adaptation. We evaluate our method using the setup in [2, 3]

Method	MNIST → USPS	USPS → MNIST
Source only (Ours)	0.60	0.68
DSN [2]	0.91	–
PixelDA [3]	0.96	–
SimNet [24]	0.96	0.96
M-ADDA (Ours)	**0.98**	**0.97**

We also use similar architectures for our models as those in [36]. The encoder module is the modified LeNet architecture provided in the Caffe source code [19]. The decoder is a simple linear model that transforms the encoded features into 256-unit embedding vectors. The discriminator consists of three fully connected layers: two layers with 500 hidden units followed by the final discriminator output. Each of the 500-unit layers uses a ReLU activation function (Fig. 9).

Table 1 shows the results of our experiments on the digits datasets. We see that our method achieves competitive results compared to previous state-of-the-art methods, ADDA [36]. This suggests that metric learning allows us to achieve good results for domain adaptation. Further, Table 2 shows the results of our experiments using the setup in [2, 3] where the full training set was used for both MNIST and USPS. We see that our method beats recent state-of-the-art methods in the USPS, MNIST domain adaptation challenge. However, it would be interesting to see the efficacy of M-ADDA in more complicated tasks such as the VisDA dataset challenge [23]. We

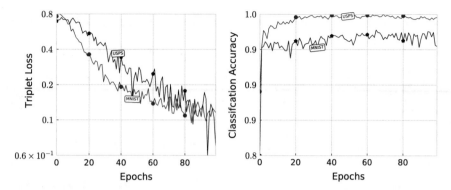

Fig. 7 Optimizing the triplet loss. (Left) The triplet loss value during the training of the source model on the USPS and MNIST datasets; (Right) The classification accuracy obtained on the target datasets

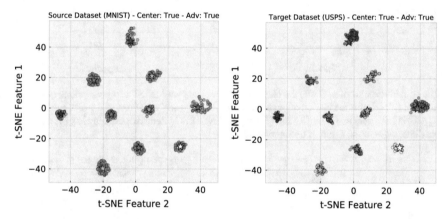

Fig. 8 M-ADDA results. (Left) the t-SNE components of the source embeddings on the MNIST dataset after training the source model. (Right) the t-SNE components of the target embeddings of the USPS dataset after training the target model. The stars represent the cluster centers of the source embeddings. The colors represent different labels

show in Fig. 7 (left) the triplet loss value during the training of the source model on the USPS and MNIST datasets. Further, Fig. 7 (right) shows the classification accuracy obtained on the target datasets with respect to the number of epochs. Higher accuracy was obtained for USPS when the model was trained on MNIST, which is expected since MNIST consists of more training examples (Fig. 8).

In Table 3, we compare between two main variations for training the target model. Center magnet only updates the target model using Eq. (4); therefore, it ignores the adversarial training part of Eq. (3). Using center magnet only to train the target model results in poor performance. This is expected since the performance highly depends on the initial clustering. We see in Fig. 10 (right) that several source cluster centers (represented as stars) contain samples corresponding to different labels. For

Table 3 **Ablation studies**. Impact of the loss terms on the classification accuracy of the target model

Method	MNIST → USPS	USPS → MNIST
Center magnet only	0.77	0.85
Adversarial adaptation only	0.93	0.92
M-ADDA	**0.98**	**0.97**

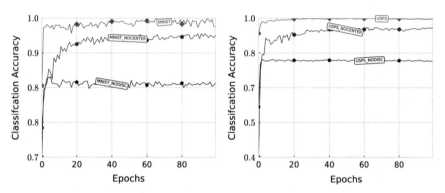

Fig. 9 **Ablation studies**. (Left) the classification accuracy on MNIST using variations of the loss function (2); (Right) the classification accuracy on USPS using variations of the loss function (2). NOCENTER refers to optimizing Eq. (3) only, and NODISC refers to optimizing Eq. (4) only. The blue lines refer to the result of optimizing Eq. (2)

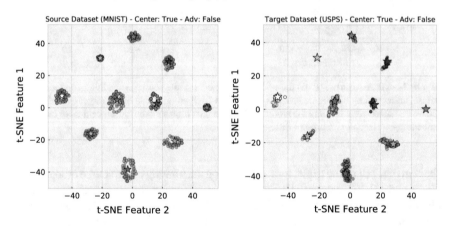

Fig. 10 **Center magnet optimization only**. The stars represent the cluster centers of the source embeddings

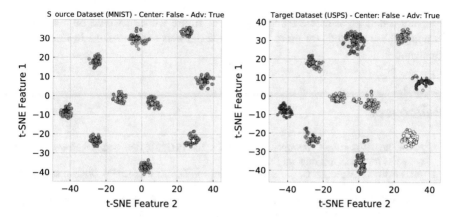

Fig. 11 Adversarial optimization only. The stars represent the cluster centers of the source embeddings

example, the samples with the pink label are clustered with those of the green label. Similarly, those with the orange label are clustered with those of the teal label. This is expected since the target model is encouraged to move the embeddings to the nearest cluster centers without having to match the extracted feature distributions between the source and target datasets (Fig. 9).

Using only the adversarial adaptation loss improves the results significantly, since having the extracted features distribution between the source and target similar is crucial. However, we see in Fig. 11 (right) that some samples are far from any cluster center which makes their class labels ambiguous, namely the pink and yellow samples that are in the center between the yellow and pink cluster centers. To address these ambiguities, the center magnet loss helps the model to regularize against them. As a result, we see in Fig. 8 (right) that better clusters are formed when we optimize the whole loss function defined in Eq. (2). This suggests that M-ADDA has strong potential in addressing the task of unsupervised domain adaptation.

5 Conclusion

We propose M-ADDA, which is a metric learning based method, to address the task of unsupervised domain adaptation. The framework consists of two main steps. First, a triplet loss is used to pretrain the source model on the source dataset. Then, we adversarially train a target model to adapt the distributions of its extracted features to match those of the source model. In parallel, we optimize a center magnet loss to regularize the output embeddings of the target model so that they form clusters that have a similar structure as that of the source model's output embeddings. We showed that this approach can perform significantly better than ADDA [36] on the digits adaptation dataset of MNIST and USPS. For future work, it would be interesting to apply these methods on more complicated datasets such as those in the VisDA challenge.

References

1. Bertinetto L, Valmadre J, Henriques JF, Vedaldi A, Torr PH (2016) Fully-convolutional Siamese networks for object tracking. In: ECCV
2. Bousmalis K, Trigeorgis G, Silberman N, Krishnan D, Erhan D (2016) Domain separation networks. In: NIPS
3. Bousmalis K, Silberman N, Dohan D, Erhan D, Krishnan D (2017) Unsupervised pixel-level domain adaptation with generative adversarial networks. In: CVPR
4. Cao X, Wipf D, Wen F, Duan G, Sun J (2013) A practical transfer learning algorithm for face verification. In: ICCV
5. Deselaers T, Alexe B, Ferrari, V (2012) Weakly supervised localization and learning with generic knowledges. IJCV
6. Finn C, Abbeel P, Levine S (2017) Model-agnostic meta-learning for fast adaptation of deep networks
7. French G, Mackiewicz M, Fisher M (2018) Self-ensembling for visual domain adaptation. In: ICLR
8. Ganin Y, Lempitsky V (2014) Unsupervised domain adaptation by backpropagation
9. Ganin Y, Ustinova E, Ajakan H, Germain P, Larochelle H, Laviolette F, Marchand M, Lempitsky V (2016) Domain-adversarial training of neural networks. JMLR
10. Ghifary M, Kleijn WB, Zhang M, Balduzzi D, Li W (2016) Deep reconstruction-classification networks for unsupervised domain adaptation. In: ECCV
11. Goodfellow I, Pouget-Abadie J, Mirza M, Xu B, Warde-Farley D, Ozair S, Courville A, Bengio Y (2014) Generative adversarial nets. In: NIPS
12. Han EHS, Karypis, G, Kumar V (2001) Text categorization using weight adjusted k-nearest neighbor classification. In: PAKDD
13. Hoffer E, Ailon N (2015) Deep metric learning using triplet network. International workshop on similarity-based pattern recognition
14. Hsu YC, Lv Z, Kira Z (2017) Learning to cluster in order to transfer across domains and tasks
15. Krizhevsky A, Sutskever I, Hinton GE (2012) Imagenet classification with deep convolutional neural networks. In: NIPS
16. Laine S, Aila T (2016) Temporal ensembling for semi-supervised learning
17. Le Cun Y, Jackel L, Boser B, Denker J, Graf H, Guyon I, Henderson D, Howard R, Hubbard W (1989) Handwritten digit recognition: applications of neural network chips and automatic learning. IEEE Commun Mag
18. LeCun Y, The MNIST database of handwritten digits. http://yann.lecun.com/exdb/mnist/
19. LeCun Y, Bottou L, Bengio Y, Haffner P (1998) Gradient-based learning applied to document recognition. IEEE
20. Li Y, Wang N, Shi J, Liu J, Hou X (2016) Revisiting batch normalization for practical domain adaptation
21. Liu MY, Tuzel O (2016) Coupled generative adversarial networks. In: NIPS
22. Long M, Cao Y, Wang J, Jordan MI (2015) Learning transferable features with deep adaptation networks
23. Peng X, Usman B, Kaushik N, Hoffman J, Wang D, Saenko K (2017) VisDA: the visual domain adaptation challenge
24. Pinheiro PO (2017) Unsupervised domain adaptation with similarity learning
25. Russo P, Carlucci FM, Tommasi T, Caputo B (2017) From source to target and back: symmetric bi-directional adaptive GAN

26. Saito K, Ushiku Y, Harada T (2017) Asymmetric tri-training for unsupervised domain adaptation
27. Sajjadi M, Javanmardi M, Tasdizen T (2016) Regularization with stochastic transformations and perturbations for deep semi-supervised learning. In: NIPS
28. Shi Z, Siva P, Xiang T (2017) Transfer learning by ranking for weakly supervised object annotation
29. Shrivastava A, Pfister T, Tuzel O, Susskind J, Wang W, Webb R (2017) Learning from simulated and unsupervised images through adversarial training. In: CVPR
30. Song HO, Xiang Y, Jegelka S, Savarese S (2016) Deep metric learning via lifted structured feature embedding. In: CVPR
31. Sun B, Saenko K (2016) Deep coral: correlation alignment for deep domain adaptation. In: ECCV
32. Sun B, Feng J, Saenko K (2016) Return of frustratingly easy domain adaptation. In: AAAI
33. Tarvainen A, Valpola H (2017) Mean teachers are better role models: weight-averaged consistency targets improve semi-supervised deep learning results. In: NIPS
34. Tzeng E, Hoffman J, Zhang N, Saenko K, Darrell T (2014) Deep domain confusion: maximizing for domain invariance
35. Tzeng E, Hoffman J, Darrell T, Saenko K (2015) Simultaneous deep transfer across domains and tasks. In: ICCV
36. Tzeng E, Hoffman J, Saenko K, Darrell T (2017) Adversarial discriminative domain adaptation
37. Tzeng E, Hoffman J, Saenko K, Darrell T (2017) Adversarial discriminative domain adaptation. In: CVPR
38. Wang M, Deng W (2018) Deep visual domain adaptation: a survey. Neurocomputing
39. Weinberger KQ, Saul LK (2009) Distance metric learning for large margin nearest neighbor classification. JMLR
40. Xing EP, Jordan MI, Russell SJ, Ng AY (2003) Distance metric learning with application to clustering with side-information. In: NIPS

XGAN: Unsupervised Image-to-Image Translation for Many-to-Many Mappings

Amélie Royer⊙, Konstantinos Bousmalis, Stephan Gouws, Fred Bertsch, Inbar Mosseri, Forrester Cole and Kevin Murphy

Abstract Image translation refers to the task of mapping images from a visual domain to another. Given two unpaired collections of images, we aim to learn a mapping between the corpus-level style of each collection, while preserving semantic content shared across the two domains. We introduce XGAN, a dual adversarial auto-encoder, which captures a shared representation of the common domain semantic content in an unsupervised way, while jointly learning the domain-to-domain image translations in both directions. We exploit ideas from the domain adaptation literature and define a *semantic consistency loss* which encourages the learned embedding to preserve semantics shared across domains. We report promising qualitative results for the task of face-to-cartoon translation. The cartoon dataset we collected for this purpose, "CartoonSet", is also publicly available as a new benchmark for semantic style transfer at https://google.github.io/cartoonset/index.html.

Keywords Generative models · Style transfer · Domain adaptation

A. Royer—Work done while at Google Brain London, UK.

A. Royer (✉)
IST Austria, 3400 Klosterneuburg, Austria
e-mail: aroyer@ist.ac.at

K. Bousmalis · S. Gouws
Google Brain, London, UK
e-mail: konstantinos@google.com

S. Gouws
e-mail: sgouws@google.com

K. Bousmalis
Deepmind, London, UK

F. Bertsch
Google Brain, Mountain View, CA, USA

I. Mosseri · F. Cole
Google Research, Cambridge, MA, USA

K. Murphy
Google Research, Mountain View, CA, USA

© Springer Nature Switzerland AG 2020
R. Singh et al. (eds.), *Domain Adaptation for Visual Understanding*,
https://doi.org/10.1007/978-3-030-30671-7_3

1 Introduction

Image-to-image translation—learning to map images from one domain to another—covers several classical computer vision tasks such as style transfer (rendering an image in the style of a given input [4]), colorization (mapping grayscale images to color images [26]), super-resolution (increasing the resolution of an input image [13]), or semantic segmentation (inferring pixel-wise semantic labeling of a scene [18]). Learning such mappings requires an underlying understanding of the shared information between the two domains. In many cases, supervision encapsulates this knowledge in the form of labels or paired samples. This holds, for instance, for colorization, where ground-truth pairs are easily obtained by generating grayscale images from colored inputs.

In this work, we consider the task of *unsupervised semantic style transfer*: learning to map an image from one domain into the style of another domain without altering its semantic content (see Fig. 1). In particular, we experiment on the task of translating faces to cartoons. Note that without loss of generality, a photo of a face can be mapped to many valid cartoons, and vice-versa. Semantic style transfer is, therefore, a *many-to-many mapping* problem, for which obtaining labeled examples are ambiguous and costly. Furthermore, in this unsupervised setting we do not have access to supervision on shared domain semantic content (e.g., facial attributes such as hair color, eye color, etc.). Instead, we propose an encoder–decoder structure with a bottleneck embedding shared across the two domains to capture common semantics as a latent representation.

The key issue is thus to learn an embedding that preserves semantic facial attributes (hair color, eye color, etc.) between the two domains with little supervision, and to incorporate it within a generative model to produce the actual domain translations. Although this chapter specifically focuses on the face-to-cartoon setting, many other

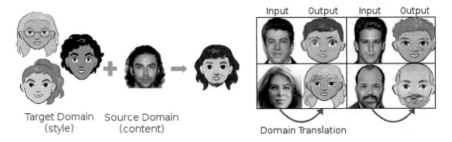

Fig. 1 Semantic style transfer is the task of adapting an image to the visual appearance of another domain without altering its semantic content given only two unpaired image collections without pairs supervision (*left*). We define semantic content as characteristic attributes which are shared across domains but do not necessarily appear the same at the pixel-level. For instance, cartoons and faces have a similar range of hair color but with very different appearances, e.g., blonde hair is bright yellow in cartoons. The proposed XGAN applied on the face-to-cartoon task yields a shared representation that preserves important face semantics such as hair style or face shape (*right*)

examples fall under this category: mapping landscape pictures to paintings (where the different scene objects and their composition describe the input semantics), transforming sketches to images, or even cross-domain tasks such as generating images from text. We only rely on two unlabeled training image collections or *corpora*, one for each domain, with no known image pairings across domains. Hence, we are faced with a double *domain shift*, first in terms of global domain appearance, and second in terms of the content distribution of the two collections.

Recent work [1, 6, 10, 25, 27] report good performance using GAN-based models for unsupervised image-to-image translation when the two input domains share similar pixel-level structure (e.g., horses and zebras) but fail for more significant domain shifts (e.g., dogs and cats). Perhaps the best known recent example is CycleGAN [27]. Given two image domains $\mathcal{D}_1$ and $\mathcal{D}_2$, the model is trained with a pixel-level *cycle-consistency loss* which ensures that the mapping $g_{1\to2}$ from $\mathcal{D}_1$ to $\mathcal{D}_2$ followed by its inverse, $g_{2\to1}$, yields the identity function; i.e., ., $g_{1\to2} \circ g_{2\to1} = id$. We argue that such a pixel-level constraint is not sufficient in our setting and that we rather need a constraint in *feature space* to allow for more permissive transformations of the pixel input. To this end, we propose XGAN ("Cross-GAN"), a dual adversarial auto-encoder which learns a shared semantic representation of the two input domains in an unsupervised way, while jointly learning both domain-to-domain translations. More specifically, the domain translation $g_{1\to2}$ consists of an encoder e_1 taking inputs in $\mathcal{D}_1$, followed by a decoder d_2 with outputs in $\mathcal{D}_2$ (and likewise for $g_{2\to1}$) such that e_1 and e_2, as well as d_1 and d_2, are partially shared across domains.

The main novelty lies in how we constrain the shared embedding using techniques from the domain adaptation literature, as well as a novel *semantic consistency loss*. The latter ensures that the domain-to-domain translations preserve the semantic representation, i.e., ., that $e_1 \approx e_2 \circ g_{1\to2}$ and $e_2 \approx e_1 \circ g_{2\to1}$. Therefore, it acts as a form of self-supervision which alleviates the need for paired examples and preserves semantic feature-level information rather than pixel-level content. In the following section, we review relevant recent work before discussing the XGAN model in more detail in Sect. 3. In Sect. 4, we introduce CARTOONSET, our dataset of cartoon faces for research on semantic style transfer. Finally, in Sect. 5 we report experimental results of XGAN on the face-to-cartoon task.

2 Related Work

Recent literature suggests two main directions for tackling the semantic style transfer task: traditional style transfer and pixel-level domain adaptation. The first approach is inadequate as it only transfers texture information from a single style image, and therefore does not capture the style of an entire corpus. The latter category also fails in practice as it explicitly enforces pixel-level similarity which does not allow for significant structural change of the input. Instead, we draw inspiration from the domain adaptation and feature-level image-to-image translation literature.

Style Transfer. Neural style transfer refers to the task of transferring the texture of a *specific* style image while preserving the pixel-level structure of an input content image [4, 9]. Recently, [14, 15] proposed to instead use a dense local patch-based matching approach in the feature space, as opposed to global feature matching, allowing for convincing transformations between visually dissimilar domains. Still, these models only perform image-specific transfer rather than learning a global *corpus-level* style and do not provide a meaningful shared domain representation. Furthermore, the generated images are usually very close to the original input in terms of pixel structure (e.g., edges) which is not suitable for drastic transformations such as face-to-cartoon.

Domain adaptation. XGAN relies on learning a shared feature representation of both domains in an unsupervised setting to capture semantic rather than pixel information. For this purpose, we make use of the domain-adversarial training scheme [3]. Moreover, recent domain adaptation work [1, 2, 22] can be framed as semantic style transfer as they tackle the problem of mapping synthetic images, easy to generate, to natural images, which are more difficult to obtain. The generated samples are then used to train a model later applied to natural images. Contrary to our work, however, they only consider pixel-level transformations.

Unsupervised Image-to-Image translation. Recent work [6, 10, 25, 27] tackle the unsupervised pixel-level image-to-image translation task by learning both cross-domain mappings jointly, each as a separate generative adversarial network, via a cycle-consistency loss which ensures that applying each mapping followed by its reverse yields the identity function. This intuitive form of self-supervision leads to good results for pixel-level transformations but often fails to capture significant structural changes [27]. In comparison, our proposed semantic consistency loss acts at the feature-level, allowing for more flexible transformations.

Orthogonal to this line of work is UNIT [7, 16, 19]. This model consists of a coupled VAEGAN architecture [12, 17] with a shared embedding bottleneck, trained with pixel-level cycle-consistency. Similar to XGAN, it learns a joint *feature-level* representation of the two domains, however, UNIT assumes that sharing high-level layers in the architecture is a sufficient constraint, while XGAN's objective explicitly introduces the semantic consistency component.

Finally, the *Domain Transfer Network* (DTN) [23, 24] is closest to our work in terms of objective and applications. The DTN architecture is a single auto-encoder trained to map images from a source to a target domain with self-supervised semantic consistency feedback. It was also successfully applied to the problem of feature-level image-to-image translation, in particular to the face-to-cartoon problem. Contrary to XGAN however, the DTN encoder is pretrained and fixed, and is assumed to produce meaningful embeddings for both the face and the cartoon domains. This assumption is very restrictive, as off-the-shelf models pretrained on natural images do not usually generalize well to other domains. In fact, we show in Sect. 5 that a fixed encoder does not generalize well in the presence of a large domain shift between the two domains.

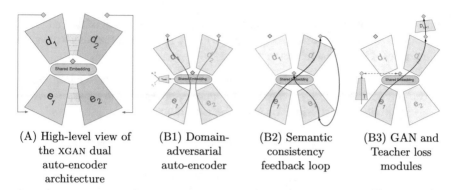

(A) High-level view of the XGAN dual auto-encoder architecture

(B1) Domain-adversarial auto-encoder

(B2) Semantic consistency feedback loop

(B3) GAN and Teacher loss modules

Fig. 2 The XGAN (**A**) objective encourages the model to learn a meaningful joint embedding (**B1**) ($\mathcal{L}_{rec}$, and $\mathcal{L}_{dann}$), which should be preserved through domain translation (**B2**) ($\mathcal{L}_{sem}$), while producing output samples of good quality (**B3**) ($\mathcal{L}_{gan}$ and $\mathcal{L}_{teach}$)

3 Proposed Model: XGAN

Let $\mathcal{D}_1$ and $\mathcal{D}_2$ be two domains that differ in terms of *visual appearance* but share common *semantic content*. It is often easier to think of domain semantics as a high-level notion, e.g., semantic attributes, however, we do not require such annotations in practice, but instead consider learning a feature-level representation that automatically captures these shared semantics. Our goal is thus to learn in an unsupervised fashion, i.e., ., without paired examples, a joint domain-invariant embedding: semantically similar inputs across domains will be embedded nearby in the learned feature space.

Architecture-wise, XGAN is a dual auto-encoder on domains $\mathcal{D}_1$ and $\mathcal{D}_2$ Fig. 2A. We denote by e_1 the encoder and by d_1 the decoder for domain $\mathcal{D}_1$; likewise e_2 and d_2 for $\mathcal{D}_2$. For simplicity, we also denote by $g_{1\to2} = d_2 \circ e_1$ the transformation from $\mathcal{D}_1$ to $\mathcal{D}_2$; likewise $g_{2\to1}$ for $\mathcal{D}_2$ to $\mathcal{D}_1$.

The training objective can be decomposed into five main components: the *reconstruction* loss, $\mathcal{L}_{rec}$, encourages the learned embedding to encode meaningful knowledge for each domain; the *domain-adversarial* loss, $\mathcal{L}_{dann}$, pushes embeddings from $\mathcal{D}_1$ and $\mathcal{D}_2$ to lie in the same subspace, bridging the domain gap at the semantic level; the *semantic consistency* loss, $\mathcal{L}_{sem}$, ensures that input semantics are preserved after domain translation; $\mathcal{L}_{gan}$ is a simple generative adversarial (GAN) objective, encouraging the model to generate more realistic samples, and finally, $\mathcal{L}_{teach}$ is an optional teacher loss that distils prior knowledge from a fixed pretrained teacher embedding, when available. The total loss function is defined as a weighted sum over these five loss terms:

$$\mathcal{L}_{\text{XGAN}} = \mathcal{L}_{rec} + \omega_d \mathcal{L}_{dann} + \omega_s \mathcal{L}_{sem} + \omega_g \mathcal{L}_{gan} + \omega_t \mathcal{L}_{teach},$$

where the ω hyperparameters control the contributions from each of the individual objectives. An overview of the model is given in Fig. 2, and we discuss each objective in more detail in the rest of this section.

Reconstruction loss, $\mathcal{L}_{rec}$. $\mathcal{L}_{rec}$ encourages the model to encode enough information on each domain to perfectly reconstruct the input. More specifically $\mathcal{L}_{rec} = \mathcal{L}_{rec,1} + \mathcal{L}_{rec,2}$ is the sum of reconstruction losses for each domain.

$$\mathcal{L}_{rec,1} = \mathbb{E}_{\mathbf{x} \sim p_{\mathcal{D}_1}} \left(\|\mathbf{x} - d_1(e_1(\mathbf{x}))\|_2 \right), \text{ likewise for domain } \mathcal{D}_2 \qquad (1)$$

Domain-adversarial loss, $\mathcal{L}_{dann}$. $\mathcal{L}_{dann}$ is the domain-adversarial loss between $\mathcal{D}_1$ and $\mathcal{D}_2$, as introduced in [3]. It encourages the embeddings learned by e_1 and e_2 to lie in the same subspace. In particular, it guarantees the soundness of the cross-domain transformations $g_{1 \to 2}$ and $g_{2 \to 1}$. More formally, this is achieved by training a binary classifier, c_{dann}, on top of the embedding layer to categorize encoded images from *both* domains as coming from either $\mathcal{D}_1$ or $\mathcal{D}_2$ (see Fig. 2B1). c_{dann} is trained to maximize its classification accuracy while the encoders e_1 and e_2 simultaneously strive to minimize it, i.e., ., to confuse the domain-adversarial classifier. Denoting model parameters by θ and a classification loss function by ℓ (e.g., cross-entropy), we optimize

$$\min_{\theta_{e_1}, \theta_{e_2}} \max_{\theta_{dann}} \mathcal{L}_{dann}, \text{ where} \qquad (2)$$

$$\mathcal{L}_{dann} = \mathbb{E}_{p_{\mathcal{D}_1}} \ell(1, c_{dann}(e_1(\mathbf{x}))) + \mathbb{E}_{p_{\mathcal{D}_2}} \ell(2, c_{dann}(e_2(\mathbf{x})))$$

Semantic consistency loss, $\mathcal{L}_{sem}$. Our key contribution is a semantic consistency feedback loop that acts as self-supervision for the cross-domain translations $g_{1 \to 2}$ and $g_{2 \to 1}$. Intuitively, we want the semantics of input $\mathbf{x} \in \mathcal{D}_1$ to be preserved when translated to the other domain, $g_{1 \to 2}(\mathbf{x}) \in \mathcal{D}_2$, and similarly for the reverse mapping. However, this consistency property is hard to assess at the pixel-level as we do not have paired data and pixel-level metrics are suboptimal for image comparison. Instead, we introduce a feature-level semantic consistency loss, which encourages the network to preserve the learned embedding during domain translation. Formally, $\mathcal{L}_{sem} = \mathcal{L}_{sem,1 \to 2} + \mathcal{L}_{sem,2 \to 1}$, where

$$\mathcal{L}_{sem,1 \to 2} = \mathbb{E}_{\mathbf{x} \sim p_{\mathcal{D}_1}} \|e_1(\mathbf{x}) - e_2(g_{1 \to 2}(\mathbf{x}))\|, \text{ likewise for } \mathcal{L}_{sem,2 \to 1}. \qquad (3)$$
$\| \cdot \|$ denotes a distance between vectors.

GAN objective, $\mathcal{L}_{gan}$. We find that generating realistic image transformations has a crucial positive effect for learning a joint meaningful and semantically consistent embedding as the produced samples are fed back through the encoders when computing the semantic consistency loss: making the transformed distribution $p(g_{2 \to 1}(\mathcal{D}_2))$ as close as possible to the original domain $p(\mathcal{D}_1)$ ensures that the encoder e_1 does not have to cope with an additional domain shift.

Thus, to improve sample quality, we add a generative adversarial loss [5] $\mathcal{L}_{gan} = \mathcal{L}_{gan,1\to2} + \mathcal{L}_{gan,2\to1}$, where $\mathcal{L}_{gan,1\to2}$ is a state-of-the-art GAN objective [5] where the generator $g_{1\to2}$ is paired against the discriminator $D_{1\to2}$ (and likewise for $g_{2\to1}$ and $D_{2\to1}$). In this scheme, a discriminator $D_{1\to2}$ strives to distinguish generated samples from real ones in $\mathcal{D}_2$, while the generator $g_{1\to2}$ aims to produce samples that confuse the discriminator. The formal objective is

$$\min_{\theta_{g_{1\to2}}} \max_{\theta_{D_{1\to2}}} \mathcal{L}_{gan,1\to2} \tag{4}$$

$$\mathcal{L}_{gan,1\to2} = \mathbb{E}_{\mathbf{x}\sim p_{\mathcal{D}_2}} (\log(D_{1\to2}(\mathbf{x}))) + \mathbb{E}_{\mathbf{x}\sim p_{\mathcal{D}_1}} (\log(1 - D_{1\to2}(g_{1\to2}(\mathbf{x}))))$$

Likewise, $\mathcal{L}_{gan,2\to1}$ is defined for the transformation from $\mathcal{D}_2$ to $\mathcal{D}_1$.

Note that the combination of the $\mathcal{L}_{gan}$ and $\mathcal{L}_{sem}$ objectives should subsume the role of the domain-adversarial loss $\mathcal{L}_{dann}$ in theory. However, $\mathcal{L}_{dann}$ plays an important role at the beginning of training to bring embeddings across domains closer, as the generated samples are typically poor and not yet representative of the actual input domains $\mathcal{D}_1$ and $\mathcal{D}_2$.

Teacher loss, $\mathcal{L}_{teach}$. We introduce an optional component to incorporate prior knowledge in the model when available, e.g., in a semi-supervised setting. $\mathcal{L}_{teach}$ encourages the learned embeddings to lie in a region of the subspace defined by the output representation of a pretrained teacher network, T. In other words, we distils feature-level knowledge from T and constrains the embeddings to a more meaningful sub-region, relative to the task on which T was trained. This can be seen as a form of regularization of the learned embedding. Moreover, $\mathcal{L}_{teach}$ is asymmetric by definition. It should not be used for both domains simultaneously as each term would potentially push the learned embedding in two different directions. Formally, $\mathcal{L}_{teach}$ (applied to domain $\mathcal{D}_1$) is defined as

$$\mathcal{L}_{teach} = \mathbb{E}_{\mathbf{x}\sim p_{\mathcal{D}_1}} \|T(\mathbf{x}) - e_1(\mathbf{x})\|, \tag{5}$$

where $\|\cdot\|$ is a distance between vectors.

3.1 Architecture and Training Procedure

We use a simple mirrored convolutional architecture for the auto-encoder. It consists of five convolutional blocks for each encoder, the two last ones being shared across domains, and likewise for the decoders (five deconvolutional blocks with the two first ones shared). This encourages the model to learn shared representations at different levels of the architecture rather than only in the middle layer. A more detailed description is given in Table 1. For the teacher network, we use the highest convolutional layer of FaceNet [21], a state-of-the-art face recognition model trained on natural images.

Table 1 Overview of the XGAN architecture used in practice. The encoder and decoder have the same architecture for both domains, and (//) indicates that the layer is shared across domain

Layer	Size
Inputs	64x64x3
conv1	32x32x32
conv2	16x16x64
(//) conv3	8x8x128
(//) conv4	4x4x256
(//) FC1	1x1x1024
(//) FC2	1x1x1024

Layer	Size
Inputs	1x1x1024
(//) deconv1	4x4x512
(//) deconv2	8x8x256
deconv3	16x16x128
deconv4	32x32x64
deconv5	64x64x3

Layer	Size
Inputs	64x64x3
conv1	32x32x16
conv2	16x16x32
conv3	8x8x32
conv4	4x4x32
FC1	1x1x1

(a) Encoder   (b) Decoder   (c) Discriminator

The XGAN training objective is to minimize (Eq. 1). In particular, the two adversarial losses ($\mathcal{L}_{gan}$ and $\mathcal{L}_{dann}$) lead to min-max optimization problems requiring careful optimization. For the GAN loss $\mathcal{L}_{gan}$, we use a standard adversarial training scheme [5]. Furthermore, for simplicity we only use one discriminator in practice, namely, $D_{1\rightarrow 2}$ which corresponds to the face-to-cartoon path, our target application. We first update the parameters of the generators $g_{1\rightarrow 2}$ and $g_{2\rightarrow 1}$ in one step. We then keep these fixed and update the parameters for the discriminator $D_{1\rightarrow 2}$. We iterate this alternating process throughout the training. The adversarial training scheme for $\mathcal{L}_{dann}$ can be implemented in practice by connecting the classifier c_{dann} and the embedding layer *via* a gradient reversal layer [3]: the feed-forward pass is unaffected, however, the gradient is backpropagated to the encoders with a sign-inversion representing the min-max alternation. We perform this update simultaneously when computing the generator parameters. Finally, we train the model with ADAM optimizer [11] and an initial learning rate of 1e-4.

4 The CartoonSet Dataset

Although previous work has tackled the task of transforming frontal faces to a specific cartoon style, there is currently no such dataset publicly available. For this purpose, we introduce a new dataset, CartoonSet,[1] which we release publicly to further aid research on this topic.

Each cartoon face is composed of 16 components including 12 facial attributes (e.g., facial hair, eye shape, etc), and 4 color attributes (such as skin or hair color) which are chosen from a discrete set of RGB values. The number of options per attribute category ranges from 3 to 111, for the largest category, hairstyle. Each

---

[1]CartoonSet, https://github.com/google/cartoonset.

of these components and their variation were drawn by the same artist, resulting in approximately 250 cartoon components artworks and 10^8 possible combinations. The artwork components are divided into a fixed set of layers that define a Z-ordering for rendering. For instance, face shape is defined on a layer below eyes and glasses, so that the artworks are rendered in the correct order. For instance, hairstyle needs to be defined on two layers, one behind the face and one in front. There are eight total layers: hair back, face, hair front, eyes, eyebrows, mouth, facial hair, and glasses. The mapping from attribute to artwork is also defined by the artist such that any random selection of attributes produces a visually appealing cartoon without any misaligned artwork, which sometimes involves handling interaction between attributes, e.g., the appearance of "short beard" will change depending on the face shape. For example, the proper way to display a "short beard" changes for different face shapes, which required the artist to create a "short beard" artwork for each face shape. We create the CartoonSet dataset from arbitrary cartoon faces by randomly sampling value for each attribute. We then filter out unusual hair colors (pink, green, etc) or unrealistic attribute combinations, which results in a final dataset of approximately 9, 000 cartoons. In particular, the filtering step guarantees that the dataset only contains realistic cartoons, while being completely unrelated to the source dataset.

5 Experiments

We experimentally evaluate our XGAN model on *semantic style transfer*; more specifically, on the task of converting images of frontal faces (source domain) to images of cartoon avatars (target domain) given an unpaired collection of such samples in each domain. Our source domain is composed of real-world frontal-face images from the VGG-Face dataset [20]. In particular, we use an image collection consisting of 18,054 uncropped celebrity frontal face pictures. As a preprocessing step, we align the faces based on eyes and mouth location and remove the background. The target domain is the CartoonSet dataset introduced in the previous section. Finally, we randomly select and take out 20% of the images from each dataset for testing purposes, and use the remaining 80% for training. For our experiments, we also resize all images to 64 × 64. As shown in Figs. 3 and 4, the two domains vary significantly in

Fig. 3 Random samples from our cartoon dataset, CartoonSet

Fig. 4 Random centered aligned samples from VGG-Face. We preprocess them with automatic portrait matting to avoid dealing with background noise

Fig. 5 Selected samples generated by XGAN on the VGG-Face (left) to CartoonSet (right) task. The figure reads row-wise: for each face-cartoon pair, the target image (cartoon) on the right was generated from the source image (face) on the left

appearance. In particular, cartoon faces are rather simplistic compared to real faces and do not display as much variety (e.g., noses or eyebrows only have a few shape options). Furthermore, we observe a major *content distribution shift* between the two domains due to the way we collected the data: for instance, certain hair color shades (e.g., bright red, gray) are overrepresented in the cartoon domain compared to real faces. Similarly, the cartoon dataset contains many samples with eyeglasses while the source dataset only has a few (Fig. 5).

Comparison to the DTN baseline. Our first evaluation is a qualitative comparison between the Domain Transfer Network (DTN) [23] and XGAN on the semantic style transfer problem outlined above. To the best of our knowledge, DTN is the current state of the art for semantic style transfer given unpaired image corpora from two domains with significant visual shift. In particular, DTN was also applied to the task of transferring face pictures to cartoons (bitmojis) in the original chapter.[2] Figure 6 shows the results of both DTN and XGAN applied to random VGG-Face samples from the test set to produce their cartoon counterpart. Evaluation metrics for style transfer are still an active research topic with no good unbiased solution yet. Hence, we choose optimal hyperparameters by manually evaluating the quality of resulting samples, focusing on accurate transfer of semantic attributes, similarity of the resulting sample to the target domain, and crispness of samples.

It is clear from Fig. 6 that DTN fails to capture the transformation function that semantically stylizes frontal faces to cartoons from our target domain. In contrast, XGAN is able to produce sensible cartoons both in terms of the style domain—the resulting cartoons look crisp and respect the specific CartoonSet style—and in terms of semantic similarity to the input samples from VGG-Face. There are some failure cases such as hair or skin color mismatch, which emerge from the weakly supervised nature of the task and the significant content shift between the two domains (e.g., red hair is overrepresented in the target cartoon dataset). In Fig. 5 we report selected XGAN samples that we think best illustrate its semantic consistency abilities, showing that the model learns a meaningful shared representation that preserves common face semantics. Additional random samples are also reported in Fig. 7.

---

[2]The original DTN code and dataset is not publicly available, hence, we instead report results from our implementation applied to the VGG-Face to CartoonSet setting.

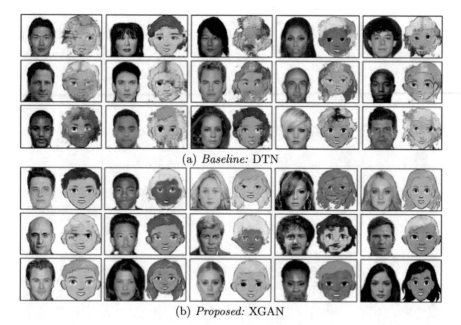

(a) *Baseline:* DTN

(b) *Proposed:* XGAN

Fig. 6 A qualitative comparison between DTN and XGAN. In both cases we present random test samples for the face-to-cartoon transformation. The tables are organized row-wise where each face input is mapped to the cartoon face immediately on its right

We believe the failure of DTN is primarily due to its assumption of a fixed joint encoder for both domains. Although the decoder learns to reconstruct inputs from the target domain almost perfectly, the semantics are not well preserved across domains and the decoder yields samples of poor quality for the domain transfer. In fact, FaceNet was originally trained on real faces inputs, hence there is no guarantee that it can produce a meaningful representation for CartoonSet samples. In contrast to our dataset, the target bitmoji domain in [23] is visually closer to real faces, as bitmojis are more realistic and customizable than the cartoon style domain we use here. This might explain the original work performance even with a fixed encoder. Our experiments suggest that using a fixed encoder is too restrictive and does not adapt well to new scenarios. We also train a DTN with a fine-tuned encoder which yields samples of better quality than the original DTN. However, this setup is very sensitive to hyperparameters choice during training and prone to mode collapse.

Comparison to CycleGAN. As we have mentioned in the related work section, Cycle-GAN [27], DiscoGAN [10], and DualGAN [25] form another family of closely related work for image-to-image translation problems. However, differently from DTN and the proposed XGAN, these models only consider a pixel-level cycle-consistency loss and do not use a shared domain embedding. Consequently, they fail to capture high-level shared semantics between significantly different domains.

(a) Source to target mapping (face-to-cartoon)

(b) Target to source mapping (cartoon-to-face)

Fig. 7 Random samples obtained by applying XGAN on faces and cartoons from the testing set for both cross-domain mappings

To explore this problem, we experiment with CycleGAN[3] on the face-to-cartoon task. We train a CycleGAN with a pix2pix [8] generator as in the original chapter, which is close to the generator we use in XGAN in terms of architecture choices and size (depth and width of the network). As shown in Fig. 8, this approach yields poor results, which is explained by the explicit pixel-level cycle-consistency loss and the fact that the pix2pix architecture contains backward connections (U-net) between the encoder and the decoder; both these features enhance pixel structure similarities which are not desirable for this task.

---

[3]CycleGAN-tensorflow, https://github.com/xhujoy/CycleGAN-tensorflow.

Fig. 8 The default CycleGAN model is not suitable for transformation between domains with very dissimilar appearances as it enforces pixel-level structural similarities

Ablation study. We conduct a number of insightful ablation experiments on XGAN. We first consider training only with the reconstruction loss $\mathcal{L}_{rec}$ and domain-adversarial loss $\mathcal{L}_{dann}$. In fact, these form the core domain adaptation component in XGAN and, as we will show, are already able to capture basic semantic knowledge across domains in practice. Second, we experiment with the semantic consistency loss and teacher loss. We show that both have complementary constraining effects on the embedding space which contributes to improving the sample consistency.

We first experiment on XGAN with only the reconstruction and domain-adversarial losses active. These components prompt the model to (i) encode enough information for each decoder to correctly reconstruct images from the corresponding domain and (ii) to ensure that the embedding lies in a common subspace for both domains. In practice in this setting, the model is robust to hyperparameter choice and does not require much tuning to converge to a good regime, i.e., ., low reconstruction error and around 50% accuracy for the domain-adversarial classifier. As a result of (ii), applying each decoder to the output of the other domain's encoder yields reasonable cross-domain translations, albeit of low quality (see Fig. 9). Furthermore, we observe that some simple semantics such as skin tone or gender are overall well preserved by the learned embedding due to the shared auto-encoder structure. For comparison, failure modes occur in extreme cases, e.g., when the model capacity is too small, in which case transferred samples are of poor quality, or when the weight ω_d is too low. In the latter case, the source and target embeddings are easily distinguishable and the cross-domain translations do not look realistic.

Second, we investigate the benefits of adding semantic consistency in XGAN via the following three components: *sharing high-level layers* in the auto-encoder leads the model to capture common semantics earlier in the architecture. In general, high-level layers in convolutional neural networks are known to encode semantic information. We performed experiments with sharing only the middle layer in the dual auto-encoder. As expected, the resulting embedding does not capture relevant shared domain semantics. Second, we use the *semantic consistency loss* as self-supervision for the learned embedding, ensuring that it is preserved through the cross-domain transformations. It also reinforces the action of the domain-adversarial loss as it constrains embeddings from the two input domains to lie close to each other. Finally, the optional *teacher loss* leads the learned source embedding to lie near the teacher

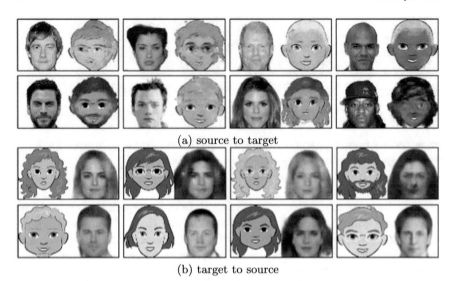

Fig. 9 Test results for XGAN with the reconstruction ($\mathcal{L}_{rec}$) and domain-adversarial ($\mathcal{L}_{dann}$) losses active only in the training objective $\mathcal{L}_{\text{XGAN}}$

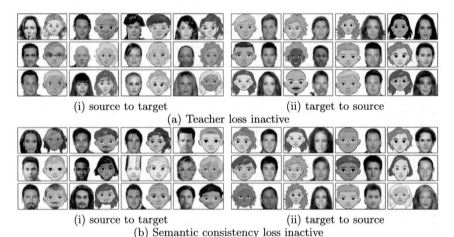

Fig. 10 Results of ablating the teacher loss ($\mathcal{L}_{teach}$) (top) and semantic consistency loss ($\mathcal{L}_{sem}$) (bottom) in the XGAN objective $\mathcal{L}_{\text{XGAN}}$

output (in our case, FaceNet's representation layer), which is meant for real faces. It acts in conjunction with the domain-adversarial loss and semantic consistency loss, whose role is to bring the source and target embedding distributions closer to each other.

In Fig. 10 we report random test samples for both domain translations when ablating the teacher loss and semantic consistency loss, respectively. While it is hard to draw conclusions from visual inspections, it seems that the teacher network has a positive regularization effect on the learned embedding by guiding it to a more realistic region: training the model without the teacher loss (Fig. 10a) yields more distorted samples, especially when the input is an outlier, e.g., person wearing a hat, or cartoons with unusual hairstyle. Conversely, when the semantic consistency is inactive (Fig. 10b), the generated samples overall display less variety. In particular, rare attributes (e.g., unusual hairstyle) are not as well preserved as when the semantic consistency term is present.

Discussions and Limitations. Our initial aim was to tackle the *semantic style transfer* problem in a fully unsupervised framework by combining techniques from domain adaptation and image-to-image translation: We first observe that using a simple setup where a partially shared dual auto-encoder is trained with reconstruction and domain-adversarial losses already suffice to produce an embedding that captures basic semantics rather well (for instance, skin tone). However, the generated samples are of poor quality and fine-grained attributes such as facial hair are not well captured. These two problems are greatly diminished after adding the GAN loss and the proposed semantic consistency loss, respectively. Failure cases still exist, especially on non-representative input samples (e.g., a person wearing a hat) which are mapped to unrealistic cartoons. Adding the teacher loss mitigates this problem by regularizing the learned embedding, however, it requires additional supervision and makes the model dependent on the specific representation provided by the teacher network.

Future work will focus on evaluating XGAN on different domain transfer tasks. In particular, though we introduced XGAN for semantic style transfer, we think the model goes beyond this scope and can be applied to classical domain adaptation problems, where quantitative evaluation becomes possible: while the pixel-level transformations are not necessary for learning the shared embedding, they are beneficial for learning a meaningful representation across visual domains, when combined with the self-supervised semantic consistency loop.

6 Conclusions

In this work, we introduced XGAN, a model for unsupervised domain translation applied to the task of semantically consistent style transfer. In particular, we argue that similar to the domain adaptation task, learning image-to-image translation between two structurally different domains requires learning a high-level joint semantic representation while discarding local pixel-level dependencies. Additionally, we proposed a semantic consistency loss acting on both domain translations as a form of self-supervision.

We reported promising experimental results on the task of face-to-cartoon that outperform the current baseline. We also showed that additional weak supervision, such as a pretrained feature representation, can easily be added to the model in the form of teacher knowledge. It acts as a good regularizer for the learned embeddings and generated samples. This is particularly useful for natural image datasets, for which off-the-shelf pretrained models are abundant.

References

1. Bousmalis K, Silberman N, Dohan D, Erhan D, Krishnan D (2017) Unsupervised pixel-level domain adaptation with generative adversarial networks. In: CVPR
2. Bousmalis K, Trigeorgis G, Silberman N, Krishnan D, Erhan D (2016) Domain separation networks. In: NIPS
3. Ganin Y, Ustinova E, Ajakan H, Germain P, Larochelle H, Laviolette F, Marchand M, Lempitsky V (2016) Domain-adversarial training of neural networks. J Mach Learn Res
4. Gatys LA, Ecker AS, Bethge M (2016) Image style transfer using convolutional neural networks. In: CVPR
5. Goodfellow I, Pouget-Abadie J, Mirza M, Xu B, Warde-Farley D, Ozair S, Courville A, Bengio Y (2014) Generative adversarial nets. In: NIPS
6. Hoffman J, Tzeng E, Park T, Zhu J, Isola P, Saenko K, Efros AA, Darrell T (2017) CyCADA: cycle-consistent adversarial domain adaptation. CoRR arXiv:abs/1711.03213
7. Huang X, Liu MY, Belongie S, Kautz J (2018) Multimodal unsupervised image-to-image translation. arXiv:1804.04732
8. Isola P, Zhu JY, Zhou T, Efros AA (2017) Image-to-image translation with conditional adversarial networks. In: CVPR
9. Johnson J, Alahi A, Fei-Fei L (2016) Perceptual losses for real-time style transfer and super-resolution. In: ECCV
10. Kim T, Cha M, Kim H, Lee JK, Kim J (2017) Learning to discover cross-domain relations with generative adversarial networks. In: ICML
11. Kingma DP, Ba J (2015) Adam: a method for stochastic optimization. In: ICLR
12. Larsen ABL, Sønderby SK, Winther O (2016) Autoencoding beyond pixels using a learned similarity metric. In: ICML
13. Ledig C, Theis L, Huszár F, Caballero J, Cunningham A, Acosta A, Aitken A, Tejani A, Totz J, Wang Z, et al (2017) Photo-realistic single image super-resolution using a generative adversarial network. In: CVPR
14. Li C, Wand, M (2016) Combining Markov random fields and convolutional neural networks for image synthesis. In: CVPR
15. Liao J, Yao Y, Yuan L, Hua G, Kang SB (2017) Visual attribute transfer through deep image analogy. ACM Trans Graph
16. Liu M, Breuel T, Kautz J (2017) Unsupervised image-to-image translation networks. In: NIPS
17. Liu MY, Tuzel O (2016) Coupled generative adversarial networks. In: NIPS
18. Long J, Shelhamer E, Darrell T (2015) Fully convolutional networks for semantic segmentation. In: CVPR
19. Ma L, Jia X, Georgoulis S, Tuytelaars T, Van Gool L (2018) Exemplar guided unsupervised image-to-image translation. arXiv:1805.11145
20. Parkhi OM, Vedaldi A, Zisserman A (2015) Deep face recognition. In: BMVC
21. Schroff F, Kalenichenko D, Philbin J (2015) FaceNet: a unified embedding for face recognition and clustering. In: CVPR
22. Shrivastava A, Pfister T, Tuzel O, Susskind J, Wang W, Webb R (2017) Learning from simulated and unsupervised images through adversarial training. In: CVPR

23. Taigman Y, Polyak A, Wolf L (2017) Unsupervised cross-domain image generation. In: ICLR
24. Wolf L, Taigman Y, Polyak A (2017) Unsupervised creation of parameterized avatars. In: ICCV
25. Yi Z, Zhang H, Tan P, Gong M (2017) DualGan: unsupervised dual learning for image-to-image translation. In: ICCV
26. Zhang R, Isola P, Efros AA (2016) Colorful image colorization. In: ECCV
27. Zhu JY, Park T, Isola P, Efros AA (2017) Unpaired image-to-image translation using cycle-consistent adversarial networks. In: ICCV

Improving Transferability of Deep Neural Networks

Parijat Dube, Bishwaranjan Bhattacharjee, Elisabeth Petit-Bois
and Matthew Hill

Abstract Learning from small amounts of labeled data is a challenge in the area of deep learning. This is currently addressed by Transfer Learning, where one learns the small dataset as a transfer task from a larger source dataset. Transfer Learning can deliver higher accuracy if the hyperparameters and source dataset are chosen well. One of the important parameters is the learning rate for the layers of the neural network. We show through experiments on the ImageNet22k and Oxford Flowers datasets that improvements in accuracy in range of 127% can be obtained by proper choice of learning rates. We also show that the images/label parameter for a dataset can potentially be used to determine optimal learning rates for the layers to get the best overall accuracy. We additionally validate this method on a sample of real-world image classification tasks from a public visual recognition API.

Keywords Deep learning · Transfer learning · Fine-tuning · Deep neural network · Experimental

E. Petit-Bois—Work done when the author was a student intern at IBM Research AI.

P. Dube (✉) · B. Bhattacharjee · E. Petit-Bois · M. Hill
IBM Research AI, Yorktown Heights, NY 10598, USA
e-mail: pdube@us.ibm.com

B. Bhattacharjee
e-mail: bhatta@us.ibm.com

E. Petit-Bois
e-mail: epetitbo@students.kennesaw.edu

M. Hill
e-mail: mh@us.ibm.com

B. Bhattacharjee
Kennesaw State University, Kennesaw, GA 30144, USA

© Springer Nature Switzerland AG 2020
R. Singh et al. (eds.), *Domain Adaptation for Visual Understanding*,
https://doi.org/10.1007/978-3-030-30671-7_4

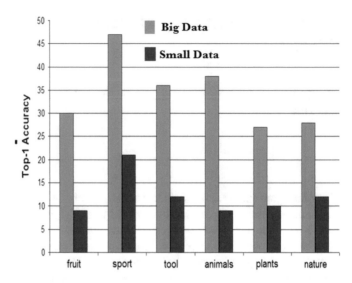

Fig. 1 Impact of data size on learning accuracy

1 Introduction

Deep Learning has become all pervasive in many application domains like Vision, Speech, and Natural Language Processing [13]. This can be partly attributed to the availability of fast processing units like GPUs as well as better neural network designs. The availability of large, open source, general-purpose labeled data has also helped the penetration of Deep Learning into these domains.

The accuracy obtained on a learning task depends on the quality and quantity of training data. As Fig. 1 shows, with larger amounts of data, for the same learning task, one can obtain much better accuracy. In this figure, the accuracy obtained on various categories of ImageNet22K [5] are shown with the big data being 10x bigger in size than the small data. While large, open source, general purpose, labeled data is available, customers often have specific needs for training. For example, a doctor may be interested in using Deep Learning for Melanoma Detection [4]. The amount of labeled data available in these specific areas is rather limited. In situations like these, the training accuracy can be negatively impacted if trained with only this limited data. To alleviate this problem, one can fall back on Transfer Learning [14, 17].

In Transfer Learning, one takes a model, trained on a potentially large dataset (called the source dataset) and then learns a new, smaller dataset (called the target dataset) as a transfer task (T) on it. This can be achieved by fine-tuning the weights of neurons in the pretrained model using the target dataset. Fine-tuning is a technique to leverage the information contained in a source dataset by tweaking the weights of its pretrained network while training the model for a target dataset. It has been shown that models trained on the source dataset learn basic concepts which will be useful in learning the target dataset [18].

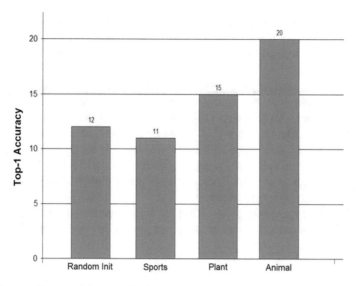

Fig. 2 Impact of base model on transfer learning accuracy

In the area of vision, the neural networks tend to be quite deep in terms of layers [10]. It has been shown that the layers learn different concepts. The initial layers learn very basic concepts like color, edges, shapes, and textures while later layers learn complex concepts [12]. The last layer tends to learn to differentiate between the labels supported by the source dataset.

The key challenges to Transfer Learning are how, what, and when to transfer [17]. One needs to address key questions like the selection of the source dataset, the neural network to use, the various hyperparameter settings as well as the type of training method to apply on the selected neural network and dataset. Figure 2 shows the accuracy obtained while training on the Tool category of ImageNet22K on models created from different source categories of ImageNet22K like Sports, Animals, Plant as well as random initialization. As the figure indicates, accuracy varied from −8% to +67% improvement over the random initialization (no Transfer Learning) case.

When performing Transfer Learning using deep learning, a popular method of training is using Stochastic Gradient Descent (SGD) [3]. In SGD, the key hyperparameters to control the descent are the block size, the step size, and the learning rate. In the case of Transfer Learning, the learning rate can be set for every layer of the neural network. This controls how much the weights in each layer change as training progresses on the target dataset. A lower learning rate for a layer allows the layer to retain what it has learned from the source data longer. Conversely, a higher learning rate forces the layer to relearn those weights quicker for the target dataset. For Transfer Learning, the concepts learned in the early layers tend to have high value since the source dataset is typically large, and the early layers represent lower level features that are transferable to the target task. If the rates are large, then the weights could change significantly and the neural network could overlearn on the target task, especially if the target task has a limited amount of training data. The accuracy that is obtained on the target task depends on the proper selection of all these parameters.

In this chapter, we study the impact of individualized layer learning rates on the accuracy of training. We use a large dataset called ImageNet22K [5] and a small dataset called the Oxford Flowers [15] for our experiments. These experiments are done on a deep residual network [10]. We show that the number of images per label plays an important role in the choice of the learning rate for a layer. We also share preliminary results on real-world image classification tasks which indicate graduated learning rates across a network, such that early layers change slowly and allow for better accuracy on the target dataset.

The chapter is organized as follows: In Sect. 2, we describe related work. In Sects. 3 and 4 we describe our experimental setup and present our results, respectively. We conclude in Sect. 5.

2 Related Work

Several approaches are proposed to deal with the problem of learning with small amounts of data. These include one-shot learning [8], zero-shot learning [16], multitask learning [1, 7], and generic transfer learning [2, 9, 18].

Multitask learning simultaneously trains the network for multiple related tasks by finding a shared feature space [1]. An example is Neural Machine Translation (NMT) where the same network is used for translation to different languages [7]. In [9] a joint fine-tuning approach is proposed to tackle the problem of training with insufficient labeled data. The basic idea is to select a subset of training data from source dataset (with similar low-level features as target dataset) and use it to augment the training dataset for target task. Here, the convolutional layers of the resulting network are fine-tuned for both the source and target tasks. Our work is targeted for scenarios where source dataset is not accessible and fine-tuning is only possible using a target dataset.

It was established in [18] that fine-tuning all the layers of the neural network gives the best accuracy. However, there is no study on the sensitivity of accuracy to the degree of fine-tuning. In [2] it is experimentally shown for one dataset that the accuracy of a (fine-tuned) model monotonically increases with increasing learning rate and then decreases, indicating existence of an optimal learning rate before overlearning happens. We studied variation in accuracy of model with learning rate used in fine-tuning for several datasets and observed non-monotone patterns.

Another popular form of Transfer Learning is by using deep feature embeddings from a neural network to drive binary Support Vector Machines (SVMs) [2, 6]. In this approach, there are as many SVMs as categories in the target dataset and each SVM learns to classify a particular label. The feature embeddings can be taken from any layer of the neural network but, in general, is taken from the penultimate layer. This is equivalent of fine-tuning with the learning rate multipliers of all the inner layers up to the penultimate layer being kept to 0 and the last layer being changed.

3 Experimental Setup

ImageNet22k contains 21841 categories spread across hierarchical categories. We extracted some of the major hierarchies like sport, garment, fungus, weapon, plant, animal, furniture, food, person, nature, music, fruit, fabric, tool, and building to form multiple sources and target domains image sets for our evaluation. Figure 3 shows the hierarchies of ImageNet22k dataset that was used and their relative sizes in terms of number of images. Figure 4 shows representative images from some of these important domains. Some of the domains like animal, plant, person, and food contain substantially more images (and labels) than categories such as weapon, tool, or sport. This skew is reflective of real-world situations and provides a natural testbed for our method when comparing training sets of different sizes.

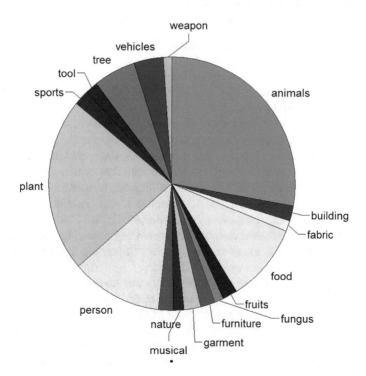

Fig. 3 Imagenet22k hierarchies used

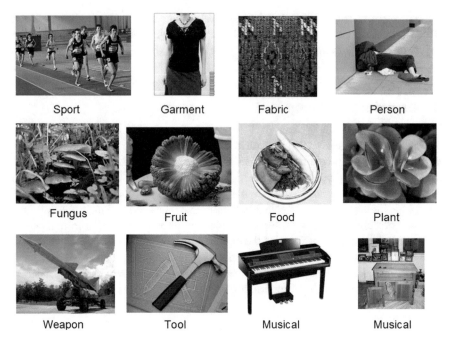

Sport	Garment	Fabric	Person
Fungus	Fruit	Food	Plant
Weapon	Tool	Musical	Musical

Fig. 4 Representative images from various Imagenet22k hierarchies used in experiments

Each of these domains was then split into four equal partitions. One was used to train the source model, two were used to validate the source and target models, and the last was used for the Transfer Learning task. One-tenth of the fourth partition was used to create a Transfer Learning target. For example, the person hierarchy has more than one million images. This was split into four equal partitions of more than 250 K each. The source model was trained with data of that size, whereas the target model was fine-tuned with one-tenth of that data size taken from one of the partitions. The smaller target datasets are reflective of real Transfer Learning tasks.

We augmented the target datasets by also using the Oxford Flower dataset [15] as a separate domain. The dataset contains 102 commonly occurring flower types with 8189 images. Out of this, a target dataset of only 10 training images per class was used. The rest of the data was used for validation.

The training of the source and target models was done using Caffe [11] and a ResNet-27 model [10]. The main components of this neural network are shown in Fig. 5. The source models were trained using SGD [3] for 900,000 iterations with a step size of 300,000 iterations and an initial learning rate of 0.01. The target models were trained with an identical network architecture, but with a training method with one-tenth of both iterations and step size. A fixed random seed was used throughout all training.

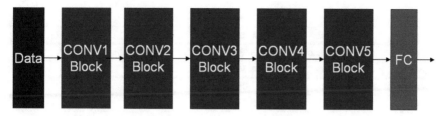

Fig. 5 Major Blocks of the ResNet model used in the experiments

4 Results and Discussion

Fine-tuning the weights involves initializing the weights to the values from the source model and then adjusting them to reduce the classification loss with the target dataset. Typically in fine-tuning a source model to a target domain, the practice is to keep the weights of all the inner layers unchanged and only fine-tune the weights of the last fully connected layer. The parameter which controls the degree of fine-tuning is the learning rate. Let $IL - n/LL - m$ be a transfer learning fine-tuning experiment where the inner layers learning rate (IL) is at n and outer layer learning rate (LL) is at m, with $n < m$. We are assuming a uniform learning rate for all the inner layers for most of the experiments. For those where the inner learning rate was varied, it is specifically mentioned in the chapter.

4.1 Fine-tuning Last Layer

We first did some experiments to quantify the gains possible by varying the learning rate of the last layer in fine-tuning while keeping all the inner layers weights unchanged. Table 1 compares the difference in accuracy of trained model for two different values of learning rate of the last layer, 0.01 and 0.1, corresponding to experiments $IL - 0/LL - 0.01$ and $IL - 0/LL - 0.1$. Observe that the accuracy is sensitive to the choice of LL and significant gains in accuracy (up to 127%) are achievable for certain domains by just choosing the best value of LL.

4.2 Fine-tuning Inner Layers

An earlier work [2, 18] has observed that fine-tuning inner layers along with the last layer can give better accuracy compared to only fine-tuning the last layer. However their observation was based on limited datasets. We are interested in studying how the accuracy changes with IL for a fixed LL with the following objectives:

Table 1 Transfer learning accuracy with varying LL

Target	Source	LL-0.01 (%)	LL-0.1 (%)	% Gain
Fabric	Garment	**13.09**	11.33	15.47
Tool	Weapon	14.54	**14.78**	1.63
Oxford	Plants	**91.06**	73.17	24.44
Food	Fruit	**5.71**	5.07	12.52
Fungus	Plant	**13.12**	5.80	127.79
Person	Food	**4.49**	2.81	59.75
Fruit	Garment	9.30	**10.50**	12.92
Music	Plant	**15.37**	9.47	62.22

(i) Identify patterns which can be used to provide guidelines for choosing LL and IL for a given source/target dataset.

(ii) Find correlation between dataset features like images per label, similarity between source and target datasets, and the choice of IL/LL.

(iii) Quantify possible gains in accuracy for different datasets by exploring the space of LL and IL values and hence establish the need to develop algorithms for identifying the right set of fine-tuning parameters for a given source/target dataset.

To this end, we conducted experiments varying IL for a fixed LL. We divided the experiments into two sets based on perceived semantic closeness of source and target domains. Set A (B) consists of experiments where the source and target datasets are semantically close (far). Thus we have,

$$A = \{fabric_t/garment_s, tool_t/weapon_s, oxford_t/plants_s,$$
$$food_t/fruit_s, fungus_t/plant_s\}, \text{ and}$$
$$B = \{person_t/food_s, fruit_t/garment_s, music_t/plant_s\}$$

Figures 6 and 7 show the accuracy obtained by increasing IL by powers of 10 between 0 and LL for $LL = 0.01$ and 0.1. So when $LL = 0.01(0.1)$, IL took values in $\{0, 0.0001, 0.001, 0.01\}(\{0, 0.0001, 0.001, 0.01, 0.1\})$.

Two patterns across different experiments are observed: (i) accuracy increases monotonically with IL and then decreases (ii) accuracy alternates between increase and decrease cycles. The variation in accuracy with IL can be significant for certain datasets. Let min_m and max_m be the minimum and maximum value of accuracy obtained when IL is varied at $LL = m$ and β_m be defined as:

$$\beta_m = \frac{max_m - min_m}{min_m} \times 100 \tag{1}$$

Fig. 6 Set A accuracy vs IL for fixed LL

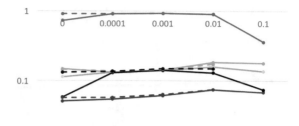

fabric_t/garment_s;0.1   tool_t/weapon_s;0.1
oxford_t/plant_s;0.1   food_t/fruit_s;0.1
fungus_t/plant_s;0.1   fabric_t/garment_s;0.01
tool_t/weapon_s;0.01   oxford_t/plant_s;0.01
food_t/fruit_s;0.01   fungus_t/plant_s;0.01

Fig. 7 Set B accuracy vs IL for fixed LL

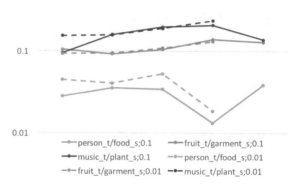

person_t/food_s;0.1   fruit_t/garment_s;0.1
music_t/plant_s;0.1   person_t/food_s;0.01
fruit_t/garment_s;0.01   music_t/plant_s;0.01

Observe that β_m represents the percentage range of possible variation in accuracy with $LL = m$ and varying IL. Figure 8 compares β_m for different datasets. All the datasets exhibit $\beta_m > 0$, with median values of $\beta_{0.01} (\beta_{0.1})$ being 28.96% (83.52%). Observe that $\beta_{0.1} > \beta_{0.01}$ for all the datasets. Also, for same dataset, the range of variation in accuracy can be quite large or small depending on LL. For example, for $oxford_t/plant_s$, $fungus_t/plant_s$, and $music_t/plant_s$ the difference $\beta_{0.1} - \beta_{0.01}$ is greater than 100 points. Thus, fine-tuning both inner and outer layers gives the best accuracy. Further the value of IL that maximizes accuracy can be different for different datasets. The pattern of variation in accuracy with IL/LL is not always monotone.

Let α_m be the value of IL that achieves the best accuracy (max_m) at $LL = m$ for a dataset. Table 2 lists α_m for different datasets. The last column in the table shows the difference $max_{0.1} - max_{0.01}$. Observe that there is no clear winner, for some datasets keeping $LL = 0.1$ and then searching for IL gives the best accuracy while for others

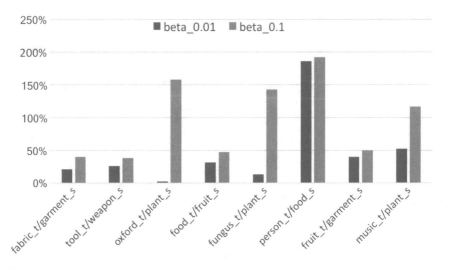

Fig. 8 Range of variation in accuracy with varying IL

Table 2 α_m for different datasets under study

Target	Source	$\alpha_{0.01}$	$\alpha_{0.1}$	$\max_{0.1} - \max_{0.01}$ (%)
Fabric	Garment	0.0001	0.0001	0.00
Tool	Weapon	0.0001	0.1	−1.41
Oxford	Plants	0.0001	0.0001	−0.88
Food	Fruit	0.01	0.01	0.98
Fungus	Plant	0.01	0.01	0.78
Person	Food	0.01	0.01	−0.71
Fruit	Garment	0.01	0.01	−0.12
Music	Plant	0.01	0.01	−2.86

$LL = 0.01$ performs better. This indicates the need for joint optimization over the space of LL and IL to get the best accuracy.

We are interested in identifying correlation between source/target dataset features and α_m. The first feature that we consider is images/label in the target dataset. Intuitively with more labeled data for the target domain, we can be more aggressive (i.e., use larger IL and LL) in fine-tuning. Figure 9 plots α_m versus images/label in target for $m = 0.01$ and 0.1. For both these cases we observe that α_m increases with images/label. However there is one anomaly, $\alpha_{0.1} = 0.1$ for $person_t/food_s$, though $person_t$ has smaller images/label. This seems to allude that other features of source/target datasets also dictate the choice of learning rates. We are currently investigating this direction with the hope to develop some functional mapping between the features of source/target datasets and α_m. This knowledge can be leveraged to develop intelligent algorithms to identify the best learning rate for inner layers and outer layers for a given source/target dataset.

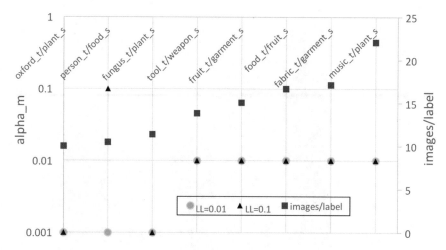

Fig. 9 Correlation between α_m and images/label

4.3 Graduated Fine-tuning of Inner Layers

We also investigated how the top-1 accuracy varies if the inner layer learning rate multipliers are not kept at a fixed value but varied. With the assumption that very basic concepts learned in the earlier layers are more important for transfer learning than later layers which map to complex concepts, we varied the learning rate multipliers in steps within the inner layers.

Oxford Flowers Dataset The ResNet-27 we are using for throughout these experiments has inner convolutional layers organized in five stages, conv1 through conv5 as shown in Fig. 5. We can denote the learning rate multiplier for each of these five stages as IL_1 through IL_5. We measured the accuracy of fine-tuning when we kept the inner learning rate multiplier ($IL_1..IL_5$) equal across stages, (at a fixed value of either 1, 2, or 5) and also compared to using a graduated set of values. In this case, each convolutional stage was assigned a multiplier (like 0, 1, 2, and 5), with conv1 and conv2 using the same (first, smallest) multiplier, and conv3, 4, and 5 using the successive, larger multipliers. (Meaning IL_1 was equal to IL_2.) In each case we set the learning rate multiplier LL of the last layer to 10. Figure 10 shows the top-1 accuracy for different IL configurations with Oxford flowers as the target dataset and plant as the source dataset with the base learning rate at 0.001. As the chart shows, the best accuracy was achieved when the learning rate multipliers were graduated.

Real-World Image Classification Tasks Next, we sought to validate these observations on training data "in the wild". IBM operates a public cloud API called Watson Visual Recognition[1] which enables users to train their own image classifier by providing labeled example images, while images provided to the API are not used to

---

[1] https://www.ibm.com/watson/developercloud/visual-recognition/api/v3.

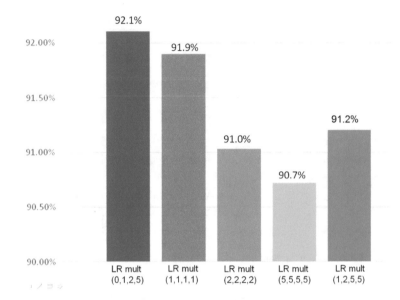

Fig. 10 Top-1 accuracy with varying inner LR mult and fixed outer LR mult at 20

train anything aside from that user's model. Users can opt-in to allow their image data to be used to help evaluate changes in the training engine. From the many training tasks that were opted-in, we took a random sample of 70 tasks. We did not manually inspect the images, but based on the names given to the labels, we presumed they represented a wide variety of image types, including industrial, consumer, scientific, and social domains as shown in Fig. 11. Based on the languages of the class labels, we had a wide geographic range as well. The average number of training images per task was about 250, with an average of 5 classes in each, so a mean of 50 image examples per class. We randomly split these into 80% for training and 20% for validation, leaving 40 training images per class on average.

For each of the 70 training tasks, we created a baseline model that was a ResNet-27 initialized with weights from an ImageNet1K model. We set the base learning rate to 0.001 and the LL to 10. The IL was set to 0. We fine-tuned the network for 20 epochs and computed top-1 accuracy on the held-out 20% of labeled data from each task. The average top-1 accuracy across the 70 tasks was 78.1%.

For the graduated IL condition, we initialized $IL_1..IL_5$ to be $\{0, 1, 2, 4, 8\}$ and LL to be 16. We then defined a set of 11 scales, $\{0.25, 0.5, 1.0, 1.5, 2, 2.5, 3, 4, 5, 7, 10\}$. The scale is a secondary learning rate multiplier. For example, the final learning rate at scale 0.5 for conv3 (IL_3) and base learning rate 0.001 would be $0.5 * 2 * 0.001 = 0.001$. The intuition is to combine the scale factors explored in Figs. 6 and 7 with the graduated values of $IL_1..IL_5$ explored in Fig. 10.

This combination of scales and learning tasks resulted in $70 * 11 = 770$ additional fine-tuning jobs, which we ran for 20 epochs each. We evaluated the top-1 accuracy for each of these jobs. We found that if we picked the individual scale which maxi-

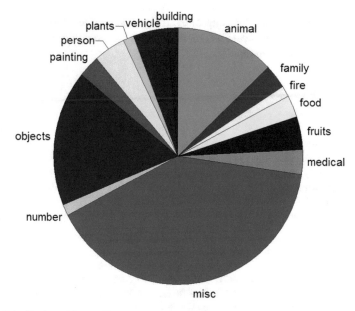

Fig. 11 Distribution of Image Classification Tasks from service API used

mized the accuracy for each job, the mean top-1 accuracy across all tasks improved from 78.1 to 88.0%, a significant gain. However, to find this maximum exhaustively requires running 11 fine-tuning jobs for each learning task. So we looked at which scale was most frequently the optimal one, and it was scale of 0.25. If we limit ourselves to one fine-tuning job per training task, and always chose this single scale, the mean top-1 accuracy across jobs had a more modest increase, from 78.1 to 79.7%.

This promising direction needs further investigation; if we could predict the optimal learning rate multiplier scale based on some known characteristic of the training task, such as number of images per class, or total number of training images, we could efficiently reach the higher accuracy point established by our exhaustive search.

5 Conclusion

Transfer Learning is a powerful method of learning from small datasets. However, the accuracy obtained from this method could vary substantially depending on the choice of the hyperparameters for training as well as the selection of the source dataset and model. We study the impact of the learning rate and multiplier which can be set for every layer of the neural network. We present experimental analysis based on the large ImageNet22K dataset, the small Oxford flower dataset and real-world image classification datsets and show that the images per label parameter could be used to determine what the learning rates. It also seems like continuously varying the learning rate for inner layers has more promise than keeping them all fixed and is a worthy direction to pursue.

References

1. Argyriou A, Evgeniou T, Pontil M (2006) Multi-task feature learning. In: Proceedings of the 19th international conference on neural information processing systems. pp 41–48
2. Bhattacharjee B, Hill M, Wu H, Chandakkar P, Smith J, Wegman M (2017) Distributed learning of deep feature embeddings for visual recognition tasks. IBM J Res Dev 61(4):1–9. https://doi.org/10.1147/JRD.2017.2706118
3. Bottou L (2010) Large-scale machine learning with stochastic gradient descent. In: Proceedings of COMPSTAT
4. Codella N, Cai J, Abedini M, Garnavi R, Halpern A, Smith JR (2015) Deep learning, sparse coding, and svm for melanoma recognition in dermoscopy images. In: Proceedings of the 6th international workshop on machine learning in medical imaging, Vol 9352. Springer, New York, pp 118–126. https://doi.org/10.1007/978-3-319-24888-2_15
5. Deng J, Dong W, Socher R, Li LJ, Li K, FeiFei L (2009) Imagenet: a large-scale hierarchical image database. In: IEEE conference on CVPR
6. Donahue J, Jia Y, Vinyals O, Hoffman J, Zhang N, Tzeng E, Darrell T (2014) Decaf: a deep convolutional activation feature for generic visual recognition. In: Proceedings of the 31st international conference on international conference on machine learning (ICML'14), Vol 32. pp I–647–I–655
7. Dong D, Wu H, He W, Yu D, Wang H (2015) Multi-task learning for multiple language translation. In: ACL
8. Fei-Fei L, Fergus R, Perona P (2006) One-shot learning of object categories. IEEE Trans Pattern Anal Mach Intell 28(4):594–611
9. Ge W, Yu Y (2017) Borrowing treasures from the wealthy: deep transfer learning through selective joint fine-tuning. In: Computer Vision and Pattern Recognition (CVPR)
10. He K, Zhang X, Ren S, Sun J (2016) Deep residual learning for image recognition. In: IEEE conference on CVPR
11. Jia Y, Shelhmer E, Donahue J, Kacayev S, long J, Girshick RB, Guadarrama S, Darrell T (2014) Caffe: convolutional architecture for fast feature embedding. In: ACM Multimedia
12. Krizhevsky A, Sutskever I, Hinton G (2012) ImageNet classification with deep convolutional neural networks. In: Neural Information Processing Systems
13. LeCun Y, Bengio Y, Hinton G (2015) Deep learning. Nature 521:436–444
14. Mou L, Meng Z, Yan R, Li G, Xu Y, Zhang L, Jin Z (2016) How transferable are neural networks in NLP applications? In: EMNLP
15. Nilsback M, Zisserman A (2008) Automated flower classification over a large number of classes. In: ICVGIP
16. Palatucci M, Pomerleau D, Hinton G, Mitchell TM (2009) Zero-shot learning with semantic output codes. In: Proceedings of the 22nd international conference on neural information processing systems. pp 1410–1418
17. Pan SJ, Yang O (2010) A survey on transfer learning. IEEE Trans Knowl Data Eng
18. Yosinski J, Clune J, Bengio Y, Lipson H (2014) How transferable are features in deep neural networks? In: Advances in Neural Information Processing Systems 27 (NIPS 2014)

Cross-Modality Video Segment Retrieval with Ensemble Learning

Xinyan Yu, Ya Zhang and Rui Zhang

Abstract Jointly modeling vision and language is a new research area which has many applications, such as video segment retrieval and video dense caption. Compared with video language retrieval, video segment retrieval is a novel task that uses natural language to retrieve a specific video segment from the whole video. One common method is to learn a similarity metric between video and language features. In this chapter, we utilize ensemble learning method to learn a video segment retrieval model. Our ensemble model aims to combine each single-stream model to learn a better similarity metric. We evaluate our method on the task of the video clip retrieval with the new proposed Distinct Describable Moments dataset. Extensive experiments have shown that our approach achieves improvement compared with the result of the state-of-art.

Keywords Video segment retrieval · Ensemble learning

1 Introduction

In the past few years, cross-modal retrieval has drawn more attention due to the rapid development of the Internet. Cross-modal retrieval is a kind of retrieval method which involves data from different modalities. It takes data from one modality as a query to retrieve data from another modality. Traditional retrieval methods only utilize single modal data. For example, if we use language query to search our interested videos on the Internet, the language query is only used to match the video caption. However, cross-modal retrieval can directly retrieve the elements in the video, such as actors,

X. Yu · Y. Zhang (✉) · R. Zhang (✉)
Cooperative Medianet Innovation Center, Shanghai Jiao Tong University,
Minhang, China
e-mail: ya_zhang@sjtu.edu.cn

X. Yu
e-mail: yuxinyan@sjtu.edu.cn

R. Zhang
e-mail: zhang_rui@sjtu.edu.cn

© Springer Nature Switzerland AG 2020   65
R. Singh et al. (eds.), *Domain Adaptation for Visual Understanding*,
https://doi.org/10.1007/978-3-030-30671-7_5

Language Query A: person begins to walk off trail
Language Query B: the child runs away from the people.

Fig. 1 Video segment retrieval is a task to retrieve a video segment from the entire video via language query. The video segment in red rectangle corresponds to the language query below. Though both language description A and B describe the same video segments at the same time, they are constructed with different words and depict the clip in different description perspectives. Description A describes the movement of the crowd in the video as a whole, and the description B depicts the movement of a specific person

actions, and objects. Therefore, cross-modal retrieval can help users to search for information in a more effective way.

In this chapter, we study a novel cross-modal retrieval task which connects video clips with natural language description. Different from traditional video language retrieval that focuses on finding the matched entire video with a given description, we want to retrieve a specific video segment from the entire video with a description. The difficulty to solve this problem is not only from the differences between each modality but also from the differences within each modality. Natural language is usually complicated and ambiguous. As shown in Fig. 1, one video segment can be described in totally different ways by two viewers. These two descriptions may be hardly considered to describe the same video scene if we only give these two sentences to another viewer. Language query A and B depict the video segment in different perspectives. Query A describes the movement of the crowd in the video while query B depicts the movement of a little child in the crowd. Although sometimes these two descriptions have the same meaning, they are not entirely made up of the same words, but of many synonyms. So it is hard to learn a suitable similarity metric to retrieve video segments with the corresponding language query.

To solve this novel and challenging problem, we utilize ensemble learning to guide the aggregation of a multi-stream cross-modal retrieval model. Ensemble learning is a widely used algorithm which combines multiple models to improve the model performance. To learn a better similarity metric for retrieval task with ensemble learning, we propose a novel method which integrates ensemble learning to guide the aggregation of multi-stream retrieval model. We conduct our experiments over the Distinct Describable Moments (DiDeMo) dataset which consists of more than 10,000 untrimmed videos with an explicit video segment caption and corresponding time stamps.

We mainly contribute in the following aspects:

- We propose a multi-stream model to retrieve the specific video segments from the entire video via text query. Multi-stream model can learn multiple common spaces for vision and language features. It could improve the learned similarity metric with ensemble learning. The combination of the ensemble model is guided by a language-based aggregation module.
- We conduct experiments on Distinct Describable Moments dataset and assess the proposed method on top-1 recall (recall@1), top-5 recall (recall@5), and mean intersection over union (mIoU). The results demonstrate that our proposed method outperforms the state-of-art.

The remainder of this chapter is structured as follows: Sect. 2 introduces related work in recent years about vision and language understanding. Section 3 gives the detail of the proposed cross-modal retrieval model. Section 4 details the experimental index, experimental setup, and experiment results. Finally, Sect. 5 concludes our work.

2 Related Work

Localizing moments in a video with natural language is a new research task which jointly models visual and language information. This task is related to both vision and language understanding.

2.1 Vision Understanding

Convolution neuron network (ConvNets) has become the most effective and widely used visual features extractor since [10] won the ImageNet Large Scale Visual Recognition Challenge (ILSVRC). Their results significantly reduced top-5 error compared with the second place. Many of the following researches [6, 18, 19] focused on improving the image recognition accuracy through increasing the depth and width of the deep network. Inspired by the success of ConvNets in the image domain, various pretrained ConvNets are transferred to extract features from the videos for video recognition. However, compared with the still image which only has appearance information, the video consists of multi-frames and has motion information between frames. Therefore, it is not suitable to directly use ConvNets trained on still image to extract video features for the lack of motion information. To integrate the motion into ConvNets, [17] used two-stream networks to model appearance and motion simultaneously. Orthogonal to the two-stream method, [21] exploited the 3D convolution kernel to concrete the spatial and temporal information across the convolution layers.

2.2   Language Understanding

Natural language processing is one of the important technologies in artificial intelligence because language is the tool for people to communicate with each other. There are also many practical natural language applications in daily life, such as semantic analysis and language translation.

Learning high-quality distributed vector representations is the most fundamental and important work in NLP task systems. Reference [13] proposed Word2Vec model to learn embedding representation. They used the correlation of source context words and the target word to model the syntactic and semantic relationship between word sequences. Due to the simple model architecture, their Continuous Bag of Words (CBOW) and Skip-gram models were efficiently trained with one trillion words. Different from the predictive-based model, GloVe [15] learned geometrical embedding vectors of words based on co-occurrence counts. This method preserved the semantic analogies and also took the corpus word occurrence statistics into consideration. To keep the ordering and semantic meaning simultaneously, [11] proposed an unsupervised learning method to learn continuous distributed vector representations for sentence and document. In this chapter, we use GloVe trained on Wikipedia corpus as our word embedding method.

2.3   Cross-Modal Understanding

Despite deep learning having been widely used and achieving success in vision and language task individually, it is still a challenge to jointly understand vision and language. Previous work has focused on tasks, such as image/video caption, image/video retrieval, and video question answering.

Early work on image caption usually used two-stage pipeline to generate sentences from still image. The semantic content is identified in the first stage and then used to generate a sentence using a language template. This two-stage pipeline simplified image caption task to only generate sentence related with some given objects and actions. Though the category of objects and actions should be elaborately selected, the limited number of categories is insufficient to model the complex sentence in the real world. Reference [24] changed this template-based model to a decoder–encoder structure. They first used deep convolution network to extract visual features from still image and then decode the fixed-length word embedding vector using Long Short-Term Memory Network (LSTM) to generate image description. Inspired by the success of this work, [23] introduced the end-to-end structure to the video caption. The difference between image caption and video caption is how to exploit temporal information of the video. To model temporal information of the video into description generation, LSTM could be used both as an encoder and decoder to generate the video caption.

Image/video-sentence retrieval is a cross-domain retrieval task. The core idea is to find the most related instance via the query from another domain. The query can be either image/video or semantic description. The common pipeline for cross-domain retrieval task is to first extract instance features from each domain and then do metric learning to narrow their similarity. Reference [3] leveraged the meaningful semantic label to improve the image classification model. They computed the similarity between joint representation of images and labels to help predict novel classes never before observed.

Reference [8] proposed the Deep Visual-Semantic Alignment (DVSA) model. They used R-CNN [5] object detector to extract image features and bidirectional LSTMs to encode sentence features. Instead of directly mapping the vision and semantic features into the common space, [9] proposed a finer-level bidirectional retrieval model that embeds the fragment of images and fragment sentences into the common space. Reference [27] integrated canonical correlation analysis (CCA) which is a traditional method for cross-modal retrieval into the deep network to match image and text. Reference [26] researched the domain structure in image–text embedding. They combined structure-preserving loss function with a bi-ranking loss to constrain the structure in each domain. Reference [12] proposed multimodal convolution network (M-CNN) to exploit the intermodal relations. They composed sentences to different-level semantic fragments to match the image. Reference [14] utilized visual and textual attention mechanisms to extract essential information from vision and language. Their dual-path attention model captured the fine-grained interplay between vision and language. Reference [22] advocated for learning a visual-semantic hierarchy over image and language.

Reference [16] collected a novel movie dataset with aligned text description— Large Scale Movie Description Challenge (LSMDC). Reference [20] studied order-embedding in joint language-visual neural network model architectures for the video text retrieval. Reference [28] proposed a high-level concept word detector and developed a semantic attention mechanism to selectively match the language description with video cue. Though many efforts have been made for video language retrieval, a few people work on localizing moments of the video via natural language query. The main obstacle for the video moment retrieval is lack of fine-grained video annotation that contains both language description and time stamps. Reference [7] collected over 10,000 unedited, personal videos and annotated video segments with referring expression. Reference [4] added sentence temporal annotations to Charades, a video dataset which consists of daily dynamic scenarios. They addressed the video segment retrieval task by using an object detection framework.

3 Methods

In this section, we introduce our multi-stream video language retrieval model and explain how to use the language information to ensemble each stream.

3.1   Model Overview

Generally, a cross-model retrieval includes two modal inputs, V and S. In our formulation, V represents the video clips and S represents the natural language. The goal of the retrieval model is to find a common embedding space for V and S. We could adopt metric learning method to learn each embedding functions F $(\cdot)$ and G $(\cdot)$. The entire cross-modal retrieval model M could be trained end-to-end with the following objective:

$$\hat{M} = \arg\min_{M} D_\theta \left(F \left(V \right), G \left(S \right) \right)$$

where D_θ $(\cdot)$ is a distance function which is used to measure the similarity between projected features of different domains. Cosine distance and Euclidean distance are two common distance functions used in the retrieval task.

In our work, we still retain the idea of projecting two domain features into the same common space. To learn a better similarity metric, we utilize the language information to aggregate the multi-stream retrieval network.

The overview of our proposed model is shown in Fig. 2. Each stream in the whole model is a basic cross-modal retrieval model which tries to project features in different domains to the same common embedding space. Then we use a language-based aggregation module to obtain the final cross-modal distance. Details of the individual modules are shown below.

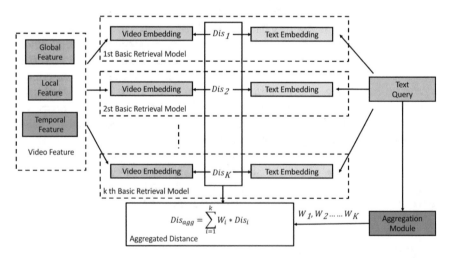

Fig. 2 The whole retrieval model contains k simple retrieval models. Video features and language query are sent into each stream to compute individual similarity distance Dis_i. The final distance is combined with k distance with aggregation module. The aggregation module exploits the semantic meaning of the query sentence to decide the importance of k basic retrieval model. Notice that our k video embedding networks share parameters of the first FC layer

3.2 Video Embedding

To localize the specific video segment from the entire video, we should take both the vision features and temporal features into consideration.

We construct our vision features using local video features V_{local} and context video features $V_{context}$. Local video features reflect what happened within a specific time span. Though the language query only depicts what occurs in the local video, context video features are important for it to provide the context information. Context information tells what happens before and after the specific time span in the video that could help localize the video segment. In our work, we first use a pretrained convolution neural network to extract features for each video frame. For a video V which consists of $[1 \dots N]$ video frames, we construct video features as

$$V_{context} = Norm_2 \left(\frac{1}{N} \sum_{i=1}^{N} Vi \right)$$

$$V_{local} = Norm_2 \left(\frac{1}{N'} \sum_{i=start}^{end} Vi \right)$$

where N represents the total number of video frames, $start$ and end represent the start and end point of the local video segment; notice that $1 \leq start < end \leq N$. We use average pooling to aggregate the features in the time span. Then, L2 Normalization after pooling is applied to rescale the vision features.

Simultaneously, putting local video features and context video features into the model could weakly help the model learn temporal relation between the video segment and the entire video. To model more temporal information that indicates whether the video segment matches the language query, we add a temporal point $[T_s, T_e]$ which represents the time span into video features. The temporal features are also normalized(to $[0, 1]$) to be in the same numerical scale with video features. Finally, we concatenate video context features $V_{context}$, video local features V_{local}, and temporal features $[T_s, T_e]$ to construct input video representation V_{input}.

Since a video consists of several still images, we could use knowledge learned from the image dataset to learn the video information. We use the model pretrained on ImageNet [10] to extract appearance feature from the video dataset. Appearance information can represent the object and other attributes in still video frames. In video recognition, motion feature is also widely used to recognize video action in the form of optical flow [17]. To model the motion information of videos, we use a video recognition network [25] to extract motion feature. In our experiments, we construct our vision features individually with the appearance and motion feature. Two ensemble retrieval models are trained respectively with appearance and motion feature and aggregated with late fusion.

The video embedding network is constructed with two fully connected layers with ReLU. The first fully connected layer in each video embedding network is shared to reduce model parameters.

3.3 Language Embedding

The natural language input is a sequence of word embedding vector representing the text query. To capture the semantic meaning of the sentence, we use the LSTM to model the query text. We first convert each word in the text query with GloVe [15] into the word embedding vector. Although the corpus which GloVe is trained on and is not related to the DiDeMo dataset, we could use GloVe as a word embedding model for its generalization. Then, the sequence of embedding vectors is put into LSTM to aggregate the semantic meaning of the sentence. Finally, the last hidden state h_t of LSTM is linear transformed with a fully connected layer to achieve embedded text features.

3.4 Language-Based Ensemble

The core problem for cross-modal retrieval is to learn a suitable similarity metric. To address this problem, we take ensemble learning into consideration. In our work, we propose a multi-stream model with a language-based ensemble. The multi-stream model contains k basic retrieval models which are shown in Fig. 1. Each basic retrieval model contains one video embedding network and one text embedding network. In our ensemble module, language query is used to aggregate the learned similarity metric in each stream. We compute the multi-stream weights with the input sentence as

$$W_i(s) = \frac{e^{p_i^{\mathrm{T}} h(s)}}{\sum_{j=1}^{k} e^{p_j^{\mathrm{T}} h(s)}} \quad i \subseteq [1 \ldots k] \tag{1}$$

where s represents the input text query, $h(\cdot)$ is the aggregate function to extract sentence meaning, and p_i denotes the linear transformer. We achieve the aggregated distance as

$$Dis_{agg} = \sum_{i=1}^{k} W_i * Dis_i \tag{2}$$

The distance Dis between the input text query and the video segment is computed in each retrieval stream first.

$$Dis = D_\theta(s, v, t) \tag{3}$$

where s is text query. t is the time stamp of the video segment v.

Our ensemble model is trained with triplet loss. Triplet loss aims to bring close the matched video clip–text pair and push away unmatched pairs. In traditional video-text retrieval task, a video–text pair is composed of video segments with its text query. Compared with that, we additionally take the time stamp of the video segment as a temporal feature. In our experiment, a training pair is denoted as $< s^i, v^i, t^i >$. s^i is the text description which describes the video segment v^i. t^i is the time interval of this video segment. During training time, we sample negative training pair within the same video or from another video. According to different sample ways, we define two triplet losses: inter-video loss and intra-video loss.

Intra-video loss Localizing a video segment from an entire video is a challenging task because a queried video segment may have little difference with its context video. To distinguish a queried video segment from its context, negative pair $< s^i, v^j, t^j >$ is sampled within the same video.

Different from traditional video retrieval task which only involves video features and text features, we integrate the temporal features in our model. The temporal features depict the position of the video clip throughout the entire video. With intra-loss, we also model the relationship between temporal features and vision features. We define intra-video loss as

$$Loss_{intra} = max\left(0, m - D_\theta\left(s^i, v^j, t^j\right) + D_\theta\left(s^i, v^i, t^i\right)\right) \tag{4}$$

where v^j is any other possible video segment in the same video. t^j denotes the time point of v^j. m is the margin variable for metric learning.

Inter-video loss Compared with intra-loss, inter-video loss is proposed to match the video segments with correct semantic concepts from other videos. For this purpose, we select a negative pair which has the same time span with the positive pair. The inter-video loss is defined as

$$Loss_{inter} = max\left(0, m - D_\theta\left(s^i, v^k, t^i\right) + D_\theta\left(s^i, v^i, t^i\right)\right) \tag{5}$$

where v^k is one possible video segment in another video. Negative pair has the same temporal features t^i with the anchor video segment v^i.

Total loss consists of weighted intra-video loss and inter-video loss.

$$Loss_{all} = \lambda Loss_{intra} + (1 - \lambda) Loss_{inter} \tag{6}$$

where λ is the parameter to adjust the importance of these two losses. In our experiment, λ is set to 0.8 for the intra-difference which is more subtle than inter-difference.

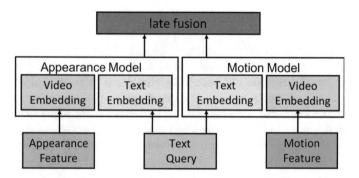

Fig. 3 The final result is obtained with aggregating the results of appearance and motion model in a late fusion way. Notice that the embedding networks in each model are trained individually

3.5 Late Fusion

For different visual input, we train two different multi-stream retrieval models individually: appearance model and motion model. The language-based aggregation module is only used to aggregate the distance computed in each single-stream model. To fuse the results of models trained with appearance and motion feature, we use the late fusion as shown in Fig. 3. Late fusion formula is defined as

$$Dis_{final} = (1 - \eta) \, Dis_{agg}^{a} + \eta Dis_{agg}^{m} \tag{7}$$

where Dis_{agg}^{a} and Dis_{agg}^{m} are the distance computed with appearance and motion model, η denotes the late fusion parameter. We set η to 0.5 via experiments on the validation set.

4 Experiments

In this section, we describe details of our training method and experiment results on the DiDeMo dataset.

4.1 Experiment Setup

We conduct experiments on the Distinct Describable Moments (DiDeMo) dataset [7]. DiDeMo consists of over 10,000 videos lasting about 25–30 s. They select about 14,000 videos from YFCC100M and eliminate those trimmed videos. The rest of the videos are then annotated by several annotators. The total number of language annotations with referring time point is over 40,000. Each description is verified to only refer to a single video moment.

The reason we choose DiDeMo as our experiment dataset is that DiDeMo contains more camera and temporal words than other video description datasets. This means the video segment in DiDeMo is depicted in multi-views. The complexity of the language description makes it more challenging to model the semantic information. It also increases retrieval difficulty that each video only has 2.57 distinct moments in average.

We report the results of our model on Rank@1 (R@1), Rank@5 (R@5), and mean intersection over union (mIoU). Each video in DeDiMo is separated into several 5 s video clips. For example, a 30 s video is broken into six five-second video segments. These video segments build up 21 possible video segments according to different time points. Our model is trained to find the most relative video segment from the 21 possible video proposals via the text query. For there are four time annotations for each video segment, four-choose-three combination is used to find the highest score.

4.2 Implement Details

Details of our training procedure are given below:

Data preprocessing We use GloVe [15] pretrained on the corpus from Wikipedia as our word embedding method. The dimension of the embedding vector is 300. As for visual features, the appearance feature is extracted from fc_7 using VGG [18] pretrained on ImageNet [2]. We also use a video recognition network [25] to extract motion feature. The two kinds of vision features could capture the video features in different views. To speed up the model training, all these features have been extracted before. The fine-tuning of the features extraction model is not implemented in our model. Two ensemble retrieval models are trained respectively with appearance and motion feature and aggregated with late fusion. These two models are denoted as appearance model and motion model corresponding to their video feature composition.

Training details We train the entire retrieval model which contains k basic model with TensorFlow [1]. k is set to 4 in our experiments. For each single-stream retrieval model, we set their hyperparameters to the same. The LSTM hidden dimension is 1000. Common embedding space is a 100-dimension vector space. The margin m in the ranking loss function is 0.1. To optimize the whole retrieval model, we apply stochastic gradient descent (SGD) to minimize the loss function.

It is insufficient to only use the aggregated loss computed by language-based aggregation module to optimize all k retrieval models. We also train all k retrieval models with ranking loss computed in each stream. The final loss function we use is

$$Loss_{final} = \alpha \sum_{i=1}^{k} Loss_{stream}^{i} + \beta Loss_{ens} \tag{8}$$

where $Loss_{stream}^{i}$ represents the loss in every single stream and is only backpropagated to each stream. $Loss_{ens}$ represents the loss computed with aggregated distance. α and β are two scalar parameters to balance the loss. In our experiment, α and β are set to 0.5 and 1.0.

4.3   Result

In this part, we evaluate our proposed multi-stream language aggregation retrieval model on the Distinct Describe Moments dataset and report the results on Rank@1, Rank@5, and mIoU. The results of our model and baseline model are shown in Table 1. We compare our model with the traditional method CCA and MCN [7].

We notice that CCA performs not as well as other methods. It is a traditional method to bridge the gap between different domains. The reason for its poor result is mainly for it cannot distinguish the subtle difference between video segment and its context. Appearance model in Table 1 represents our multi-steam retrieval model which only uses appearance feature as input. It outperforms CCA in Rank@1 and Rank@5 with 4.54 and 23.61%, but gets a lower result in mIoU. Compared with the appearance model, motion model achieves a better result on all the metric: Rank@1 = 27.78%, Rank@5 = 76.82%, and mIoU = 40.67%. This suggests that the motion feature is important in video tasks. Its better performance also attributes to the motion feature is extracted with video recognition network.

Our late fusion model achieves the best results: Rank@ = 1:29.39%, Rank@5 = 79.28%, and mIoU = 42.82%. Compared with MCN [7] which only uses single-stream retrieval model, our model leverages the language query information to aggregate the learned similarity metrics of multi-stream network. The late fusion model outperforms their results on all three evaluation metrics, respectively. The results show that our multi-stream retrieval network aggregated with language information learns a better similarity metric compared with single-stream network.

Table 1 Comparison of different methods of DiDeMo

Method	Rank@1	Rank@5	mIoU
CCA	18.11	52.11	37.82
MCN [7]	28.10	78.21	41.08
Appearance model	22.65	75.70	33.69
Motion model	27.78	76.82	40.67
Fusion model	**29.39**	**79.28**	**42.82**

Table 2 Comparison of different ensemble methods

Ensemble method	Rank@1	Rank@5	mIoU
Linear ensemble	27.01	76.73	39.62
Ours	27.78	76.82	40.67

Our language-based aggregation module unites each stream model in the spirit of ensemble learning. In our experiments, we train our aggregation module with a text query in an end-to-end way. To better analyze the effect of our text embedding module, we train a new motion model with another ensemble method. In this ensemble method, we obtain the final distance by directly inputting distance of each stream to a fully connected layer. Weights for each stream are trained as parameters of this FC layer. This ensemble method is denoted as a linear ensemble in our experiment. All the hyperparameters and optimization methods in this model are set to be the same with our standard motion model. Difference between these two models is only in the ensemble module. We compare the results of these two motion models with different ensemble methods in Table 2. Compared with the linear ensemble method, our ensemble method achieves better results on all three evaluation metrics. It demonstrates that it is better to use text information to the aggregate distance in each stream network.

5 Conclusion

In this chapter, we address the problem of localizing video segments via language query. Different from retrieving video from a video library, retrieving video segments should distinguish the subtle difference between corresponding video segments and other possible video segments within the same video. With a single-stream retrieval model, it is insufficient to learn a suitable similarity metric for this novel retrieval task. We propose multi-steam language aggregation retrieval model, in which semantic information is used to guide the aggregation of every single stream. With the language-based aggregation module, each single-stream network can be trained to obtain a better similarity metric. The whole retrieval model is optimized with in-stream loss and aggregated loss.

Our method outperforms other results on the DiDeMo dataset. Extensive experiments show that under our proposed aggregation module, multi-stream retrieval model can be effectively combined to accurately measure the distance between video and text domain. Future work will focus on excavating more video information and combining appearance and motion feature in a more efficient way.

References

1. Abadi M, Barham P, Chen J, Chen Z, Davis A, Dean J, Devin M, Ghemawat S, Irving G, Isard M et al (2016) Tensorflow: A system for large-scale machine learning
2. Deng J, Dong W, Socher R, Li LJ, Li K, Fei-Fei L (2009) ImageNet: a large-scale hierarchical image database. In: CVPR09
3. Frome A, Corrado GS, Shlens J, Bengio S, Dean J, Mikolov T et al (2013) Devise: a deep visual-semantic embedding model. In: Advances in neural information processing systems, pp 2121–2129
4. Gao J, Sun C, Yang Z, Nevatia R (2017) Tall: temporal activity localization via language query
5. Girshick R (2015) Fast r-cnn. arXiv:1504.08083
6. He K, Zhang X, Ren S, Sun J (2016) Deep residual learning for image recognition. In: Proceedings of the IEEE conference on computer vision and pattern recognition, pp 770–778
7. Hendricks LA, Wang O, Shechtman E, Sivic J, Darrell T, Russell B (2017) Localizing moments in video with natural language. arXiv:1708.01641
8. Karpathy A, Fei-Fei L (2015) Deep visual-semantic alignments for generating image descriptions. In: Proceedings of the IEEE conference on computer vision and pattern recognition, pp 3128–3137
9. Karpathy A, Joulin A, Fei-Fei LF (2014) Deep fragment embeddings for bidirectional image sentence mapping. In: Ghahramani Z, Welling M, Cortes C, Lawrence ND, Weinberger KQ (eds) Advances in neural information processing systems 27. Curran Associates, Inc, pp 1889–1897. http://papers.nips.cc/paper/5281-deep-fragment-embeddings-for-bidirectional-image-sentence-mapping.pdf
10. Krizhevsky A, Sutskever I, Hinton GE (2012) Imagenet classification with deep convolutional neural networks. In: Advances in neural information processing systems, pp 1097–1105
11. Le Q, Mikolov T (2014) Distributed representations of sentences and documents. In: International conference on machine learning, pp 1188–1196
12. Ma L, Lu Z, Shang L, Li H (2015) Multimodal convolutional neural networks for matching image and sentence. In: Proceedings of the IEEE international conference on computer vision, pp 2623–2631
13. Mikolov T, Chen K, Corrado G, Dean J (2013) Efficient estimation of word representations in vector space. arXiv:1301.3781
14. Nam H, Ha JW, Kim J (2016) Dual attention networks for multimodal reasoning and matching. arXiv:1611.00471
15. Pennington J, Socher R, Manning C (2014) Glove: global vectors for word representation. In: Proceedings of the 2014 conference on empirical methods in natural language processing (EMNLP), pp 1532–1543
16. Rohrbach A, Torabi A, Rohrbach M, Tandon N, Pal C, Larochelle H, Courville A, Schiele B (2017) Movie description. Int J Comput Vis
17. Simonyan K, Zisserman A (2014) Two-stream convolutional networks for action recognition in videos. In: Advances in neural information processing systems, pp 568–576
18. Simonyan K, Zisserman A (2014) Very deep convolutional networks for large-scale image recognition. arXiv:1409.1556
19. Szegedy C, Liu W, Jia Y, Sermanet P, Reed S, Anguelov D, Erhan D, Vanhoucke V, Rabinovich A et al (2015) Going deeper with convolutions. In: CVPR
20. Torabi A, Tandon N, Sigal L (2016) Learning language-visual embedding for movie understanding with natural-language. arXiv:1609.08124
21. Tran D, Bourdev L, Fergus R, Torresani L, Paluri M (2015) Learning spatiotemporal features with 3d convolutional networks. In: 2015 IEEE international conference on computer vision (ICCV). IEEE, pp 4489–4497
22. Vendrov I, Kiros R, Fidler S, Urtasun R (2015) Order-embeddings of images and language. arXiv:1511.06361
23. Venugopalan S, Rohrbach M, Donahue J, Mooney R, Darrell T, Saenko K (2015) Sequence to sequence - video to text. In: The IEEE international conference on computer vision (ICCV)

24. Vinyals O, Toshev A, Bengio S, Erhan D (2015) Show and tell: a neural image caption generator. In: The IEEE conference on computer vision and pattern recognition (CVPR)
25. Wang L, Xiong Y, Wang Z, Qiao Y, Lin D, Tang X, Van Gool L (2016) Temporal segment networks: towards good practices for deep action recognition. In: European conference on computer vision. Springer, pp 20–36
26. Wang L, Li Y, Lazebnik S (2016) Learning deep structure-preserving image-text embeddings. In: Proceedings of the IEEE conference on computer vision and pattern recognition, pp 5005–5013
27. Yan F, Mikolajczyk K (2015) Deep correlation for matching images and text. In: 2015 IEEE conference on computer vision and pattern recognition (CVPR). IEEE, pp 3441–3450
28. Yu Y, Ko H, Choi J, Kim G (2016) End-to-end concept word detection for video captioning, retrieval, and question answering. arXiv:1610.02947

On Minimum Discrepancy Estimation for Deep Domain Adaptation

Mohammad Mahfujur Rahman, Clinton Fookes, Mahsa Baktashmotlagh
and Sridha Sridharan

Abstract In the presence of large sets of labeled data, Deep Learning (DL) has accomplished extraordinary triumphs in the avenue of computer vision, particularly in object classification and recognition tasks. However, DL cannot always perform well when the training and testing images come from different distributions or in the presence of domain shift between training and testing images. They also suffer in the absence of labeled input data. Domain adaptation (DA) methods have been proposed to make up the poor performance due to domain shift. In this chapter, we present a new unsupervised deep domain adaptation method based on the alignment of second-order statistics (covariances) as well as maximum mean discrepancy of the source and target data with a two-stream Convolutional Neural Network (CNN). We demonstrate the ability of the proposed approach to achieve state-of-the-art performance for image classification on three benchmark domain adaptation datasets: Office-31 [27], Office-Home [37] and Office-Caltech [8].

Keywords Unsupervised domain adaptation · Domain discrepancy ·
Classification · Visual adaptation · Transfer learning · Feature learning

1 Introduction

Deep Neural Networks (DNN) [16] have brought tremendous advances across many machine learning tasks and applications such as object detection [7], object recognition [15], speech recognition [2], person re-identification [13], and machine trans-

M. M. Rahman (✉) · C. Fookes · M. Baktashmotlagh · S. Sridharan
Image and Video Laboratory, Queensland University of Technology (QUT),
Brisbane, QLD, Australia
e-mail: m27.rahman@qut.edu.au

C. Fookes
e-mail: c.fookes@qut.edu.au

M. Baktashmotlagh
e-mail: m.baktashmotlagh@qut.edu.au

S. Sridharan
e-mail: s.sridharan@qut.edu.au

© Springer Nature Switzerland AG 2020   81
R. Singh et al. (eds.), *Domain Adaptation for Visual Understanding*,
https://doi.org/10.1007/978-3-030-30671-7_6

lation [33]. For example, in [9] a DNN achieves 97.84% accuracy in multi-digit number classification from street view images because of the ability of joint feature and classifier learning of the DNN. The dramatic success of large-scale image classification based on DNNs commenced in 2012. In [15], they attained the best performance in the ImageNet Large Scale Visual Recognition Challenge (ILSVRC) by developing AlexNet. These victories were achieved in part from the accessibility of large labeled datasets such as the widely used ImageNet [15]. While the introduction of such datasets have unlocked many breakthroughs, the process of obtaining such labels still remains a time consuming and manual task.

In object recognition or classification, the training images may be different than the target images due to backgrounds, camera viewpoints, object transformations, and human selection preference. When the source data and target data distributions are dissimilar, classifier's performance can be significantly impacted. In computer vision, this is generally known as dataset bias or dataset shift [18, 34]. Learning a discriminative model of different distributions of training and test data is known as domain adaptation [24, 25, 40]. The principle objective of unsupervised domain adaptation algorithms is to interface the source and target distributions by acquiring a domain constant information where the target data are used without any labels.

Recent investigations have demonstrated that deep neural networks learn more transferable components for unsupervised domain adaptation [30]. Recently, unsupervised domain adaptation methods [10, 11, 20, 21, 26, 28, 30, 32, 38] have been proposed where features are adapted by aligning the second-order statistics of the source and target data. Although [30] introduces a new loss named Correlation Alignment (CORAL) Loss, it depends on a linear transformation, and it is not an end-to-end trainable method. After feature extraction, the linear transformation is applied, and a Support Vector Machine (SVM) classifier is trained in another phase. Moreover, the features are fixed in these types of shallow domain adaptation methods. The approach in [30] is extended in [32] to incorporate the CORAL loss directly into deep neural networks. Maximum Mean Discrepancy (MMD) is another popular metric for feature adaptation. MMD-based DA techniques have achieved great success to minimize the discrepancy between source and target data. MMD can also be incorporated with deep neural networks to achieve stronger performance over conventional methods.

In our approach, we get motivation from both of the above top performing metrics and propose a new domain adaptation method which leverages the advantages of both feature adaptation metrics: CORAL and MMD. The difference between previous research and our work is that previous approaches either minimize the source and target data discrepancy using maximum mean discrepancy or second-order statistics for feature adaptation. However, in our approach, we minimize the discrepancy using both metrics (MMD and CORAL) for feature adaptation. MMD-based methods for domain adaptation utilize symmetric transformation to distributions of the source and target data whereas CORAL-based approaches apply asymmetric transformation. However, symmetric transformations neglect the dissimilarities between the source and target data. On the other hand, asymmetric transformations attempt to link the source and target domains [31]. CORAL aligns the second-order statistics that can be reconstructed utilizing all eigenvectors and eigenvalues instead of aligning only the top k eigenvectors and eigenvalues as subspace-based methods [4].

We present an assessment of our proposed deep domain adaptation by aligning covariances or second-order statistics and maximum mean discrepancy within two streams of CNN on three benchmark datasets: Office-31 [27], the recently released Office-Home [37] and Office-Caltech [8].

In summary, the contributions of this chapter are given as follows:

- We propose a novel deep neural network approach for unsupervised domain adaptation in the context of image classification in computer vision.
- The proposed deep domain adaptation architecture jointly adapts features using two popular feature adaptation metrics: MMD and CORAL.
- We report competitive accuracy compared to the state- of-the-art methods on three benchmark domain adaptation datasets for image classification. We achieve the best average image classification accuracies on three datasets compared to other state-of-the-art methods.

The rest of the chapter is organized as follows: Sect. 2 describes related research, the proposed methodology is described in Sect. 3, Sect. 4 illustrates a comprehensive evaluation, and finally, Sect. 5 concludes the chapter.

2 Related Works

There have been many domain adaptation methods [1, 20, 21, 26, 28, 32, 35, 38] proposed in recent years to solve the problem of domain bias. All the methods can be categorized into two main categories, Conventional Domain Adaptation and Deep Domain Adaptation methods. The conventional domain adaptation methods develop their model into two stages, feature extraction and classification. In the first phase, these domain adaptation methods extract features and in the second phase, a classifiers is trained to classify the objects. However, the performance of these DA methods is not satisfactory.

Obtaining the features using deep neural network even without adaptation technique outperform the conventional DA methods by large margin. However, the results achieved with the Deep Convolutional Activation Features (DeCAF) [3] even without using any adaptation technique to the target data are remarkably better than the outcomes acquired with any conventional domain adaptation methods because DNNs extract more robust features using nonlinear transform. As a result deep neural network-based domain adaptation methods are getting popular day by day.

MMD is a popular metric for measuring the distributions of source and target samples. Tzeng et al. [36] proposed the Deep Domain Confusion (DDC) domain adaptation framework based on a confusion layer for the discrepancy between source and target data. In [35], the previous work is extended by introducing soft label distribution matching loss. Long et al. [17] proposed the Domain Adaptation Network (DAN) that propose the integration of MMDs defined among several layers, including the soft prediction layer. This idea was further improved by introducing residual transfer networks [18] and Joint Adaptation Networks [19]. Venkateswara et al. [37] proposed a new Deep Hasing Network for unsupervised domain adaptation where hash codes are used to address the domain adaptation issue.

Another popular metric for feature adaptation between domains is aligning covariance or second-order statistics which is known as Correlation Alignment. In [30, 32], unsupervised deep domain adaptation techniques have been proposed where domain shift is minimized by aligning the covariances of the source and target data. The idea is similar to Deep Domain Confusion (DDC) [36] and Deep Adaptation Network (DAN) [17] except that the CORAL loss is used instead of MMD to minimize the discrepancy between source and target data. Both [30, 32] introduces a new loss named coral loss which is the distance between the second-order statistics of the source and target representations. In [14], a deep domain adaptation approach based on the mixture of alignments of second-order or higher order scatter statistics between source and target distributions has been proposed. All these methods utilized two streams of CNN where the source network and target network combined at the classifier level. Another deep domain adaptation method is Domain-Adversarial Neural Networks (DANN) [5] which introduces a new deep learning domain adaptation approach by integrating a gradient reversal layer into the standard architecture. This gradient reversal layer do not change during forwardpropagation, but during backpropagation its gradient reverse.

In our work, we adapt the features using both CORAL and MMD metric to minimize the dissimilarity between the source and target domains. CORAL is used to align the second-order statistics and MMD is used to align higher order statistics.

3 Proposed Approach

Our proposed methodology is illustrated in Fig. 1. In Our proposed method, the features of the source and target domains are jointly adapted using CORAL and MMD metrics. The source and target data use two separate CNNs. In fc7 and fc8 layers, CORAL and MMD loss layer are added to minimize the discrepancy between the source and target data. Finally, the discrepancy between source and target data is minimized by entropy minimization of the unlabeled target data.

We consider the unsupervised domain adaptation scenario where labeled source data and unlabeled target data are available. Let us consider that the source domain data samples are $D_s = \{X_i^s\}$ with available labels $L_s = \{Y_i\}$ and the target data samples are $D_t = \{X_i^t\}$ without labels. The number of source and target samples are N_s and N_t, respectively. Let the classifiers for source domain and target domain be $F_s(X_i^s)$ and $F_t(X_i^t)$, respectively. The distribution of the data of source and target domains are nonidentical, i.e., $P_s(X_i^s, Y_s) \neq P_t(X_i^t, Y_t)$. We build a deep learning architecture which aids the learning of a transfer classifiers, such as $Y = F_s(X_i^s) = F_t(X_i^t)$ to minimize the source–target discrepancy or mismatch.

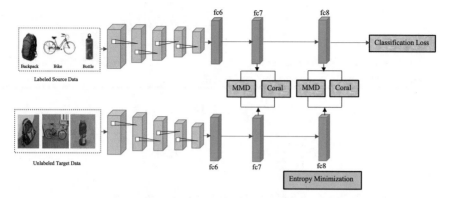

Fig. 1 Overview of our proposed methodology. The classifiers and features of both source and target data are adapted simultaneously. MMD and CORAL loss layers are added in fc7 and fc8 layers of two-stream of CNN. CORAL layer align the second-order statistics and MMD layer aligns the higher order statistics

We propose a new deep DA method which has two streams of convolutional Neural Network (CNN), one for source data and another for target data. It adapts features by aligning second-order statistics and maximum mean discrepancy of the source and target data. The discrepancy of the source and target data are minimized by the following equation:

$$\min_{F_s,F_t} D_l(D_s, D_t)_{fc7} + \min_{F_s,F_t} MMD^2(D_s, D_t)_{fc7}+$$
$$\min_{F_s,F_t} D_l(D_s, D_t)_{fc8} + \min_{F_s,F_t} MMD^2(D_s, D_t)_{fc8}+$$
$$\sum_{i=1}^{N_t} H(F_t(X_i^t)). \qquad (1)$$

Moreover, the proposed method also adapts the classifiers using entropy minimization.

The features are adapted by aligning second-order statistics as well as maximum mean discrepancy. We define the coral loss of the source and target activation features (such a loss function is used in prior work [32]) as

$$\min_{F_s,F_t} D_l(D_s, D_t) = \frac{1}{4d^2} \|C_s - C_t\|_F^2, \qquad (2)$$

where C_s and C_t denote the features covariance matrices of the source and target data and $\|.\|_F^2$ denotes the squared matrix Frobenius norm. The C_s and C_t are given by the following equation [32]:

$$C_s = \frac{1}{N_s - 1}(D_s^T D_s - \frac{1}{N_s}(1^T D_s)^T(1^T D_s)), \qquad (3)$$

$$C_t = \frac{1}{N_t - 1}(D_t^T D_t - \frac{1}{N_t}(1^T D_t)^T (1^T D_t)). \qquad (4)$$

The features are further adapted by using another popular metric for feature adaptation, MMD. The MMD loss function is defined as

$$\min_{F_s, F_t} MMD^2(D_s, D_t) =$$

$$\| \frac{1}{N_s} \sum_{i=1}^{N_s} \phi(X_i^s) - \frac{1}{N_t} \sum_{i=1}^{N_t} \phi(X_i^t) \|_H^2, \qquad (5)$$

where $\phi(X_i^s)$ denotes the feature map associated with kernel map,

$$K(X_i^s, X_i^t) = < \phi(X_i^s), \phi(X_i^t) > K(X_i^s, X_i^t). \qquad (6)$$

$K(X_i^s, X_i^t)$ is usually defined as the convex combination of L basis kernels $K_l(X_i^s, X_i^t)$ [39],

$$K(X_i^s, X_i^t) = \sum_{l=1}^{L} \beta_1 K_1(X_i^s, X_i^t) s.t. \beta_1 \geq 0, \sum_{l=1}^{L} \beta_1 = 1. \qquad (7)$$

Since feature adaptation cannot eliminate the discrepancy [18], we adapt classifiers along with feature adaptation. In this work, the classifier is adapted by decreasing the entropy of class-conditional distribution on the target data D_t (similar loss function has been proposed in prior work [18]),

$$\min_{F_t} \frac{1}{N_t} = \sum_{i=1}^{N_t} H(F_t(X_i^t)), \qquad (8)$$

where $H(\cdot)$ represents the class-conditional distribution entropy function.

3.1 Discussion

The main difference between our work and prior works is that they consider only one metric for feature adaptation whereas we consider two metrics for minimizing the discrepancy between the source and target data. In [32], CORAL layer is used in between fc8 layers of the source and target CNNs, but we used CORAL layer in between fc7 and fc8 layers. It is mentioned that the MMD metric is used in between fc8 layers in [18] and MMD layer is used in between fc6, fc7, and fc8 layers in [17]. The difference between our work and [18] is that RTN uses Residual Transfer Network and MMD metric, whereas we use simple AlexNet architecture that consists of 5 convolutional followed by three fully connected layers and CORAL and

MMD metrics to adapt the features. In our research, we have found that if multiple feature adaptation metrics are used in between fc7 and fc8, we get better accuracy using simple CNN architecture, and the best configuration of domain adaptation architecture is to use feature adaptation metric in between fc7 and fc8.

4 Experiments

In this section, we conduct extensive experiments to assess the proposed method and compare the method against recently published state-of-the-art unsupervised deep domain adaptation approaches.

4.1 Datasets

We evaluate all the methods on three standard domain adaptation benchmark datasets: Office-31 [27], Office-Home [37], and Office-Caltech [8] in the context of imagee classification.

Office-31 In the context of image classification, Office-31 is the most prominent benchmark dataset for domain adaptation. The dataset contains everyday object images from an office environment. It consists of 4110 images with 31 object categories and 3 image domains: **Amazon (A)** contains images downloaded from amazon.com, **DSLR (D)** contains images taken by Digital SLR camera and **Webcam (W)** contains images taken by web camera with different photo graphical settings. For all experiments, we use the source data with labels and target data without any labels for unsupervised domain adaptation. We conduct experiments on all six transfer tasks for all possible combinations of source and target pairs for the available three domains. The average performance of all transfer tasks is also calculated.

Office-Home The Office-Home dataset contains four domains and each domain contains images from 65 different classes (categories). The four domains are **Art (Ar), Clipart (Cl), Product (Pr)** and **Real-World (Rw)**. Art domain contains the images from sketches, paintings, ornamentation form of artistic depictions of images. Clipart domain is the collection of clipart images. The images of product domain have no background, and Real-World domain consists of images that are captured by a regular camera. It has around 15,500 images. Every category has an average of around 70 images and a maximum of 99 images. We conduct experiments on all 12 transfer tasks for all combinations of source and target pairs for the four domains. Figure 2 presents some sample images of 7 classes of Office-Home dataset.

Office-Caltech The Office-Caltech is another popular benchmark dataset in the domain adaptation community which is formed by taking the 10 common classes shared by Office-31 and Caltech-256. It has four domains named **Amazon (A), Webcam (W), DSLR (D)**, and **Caltech (C)**. We conduct experiments on all 12 transfer task as it has four different domains.

Fig. 2 There are some example images that are taken from **Office-Home** dataset. It comprises images of everyday objects. This dataset divided into four different domains; the **Clipart** domain comprises clipart images, the **Art** domain consists of sketches, paintings, artistic images, the **Product** domain comprises images which have no background and finally, the **Real-World** domain is created by taking images which are captured with a regular camera. The figure shows sample images from 7 of the 65 classes

4.2 Experimental Setup

In our method, we used two streams of Convolutional Neural Network (CNN). We extended AlexNet deep learning architecture which was pretrained on the ImageNet dataset for both streams of CNN. The dimension of the last fully connected layer (fc8) is set to the number of classes of the objects (31 for office 31, 65 for home-office and 10 for Office-Caltech datasets). We set the learning rate to 0.0001 to optimize the network. We set the batch size to 128, momentum to 0.9 and weight decay to 5×10^{-4} during training phase.

4.3 Results and Discussion

In this section, we provide the details of the performance of our method in the context of unsupervised domain adaptation where we use the labeled source data and unlabeled target data. Our proposed approach is compared with both conventional DA and recently published deep architecture-based approaches: Geodesic Flow Kernel

(GFK) [8], Transfer Component Analysis (TCA) [22], AlexNet (No adaptation) [15], VGG16 (No Adaptation) [29], Domain-Adversarial Neural Network (DANN) [5], Deep Correlation alignment (D-CORAL) [32], DAN [17], Deep Reconstruction-Classification Networks (DRCN) [6], Residual Transfer Networks (RTN) [18], and Deep Hashing Network (DAH) [37].

TCA is a traditional domain adaptation approach based on MMD-regularized Kernel primary component analysis (PCA). GFK is a subspace-based domain adaptation approach. Both TCA and GFK do not use deep neural architecture. These methods are not end-to-end approach. At first features are extracted and then the features are used in domain adaptation networks. Both AlexNet and VGG16 deep convolutional neural networks are also used as deep feature extractors without adaptation techniques to show that a standalone deep architecture works better than conventional domain adaptation techniques. DANN introduces a deep learning approach domain adaptation technique by integrating a gradient reversal layer into the standard architecture. D-CORAL is also another deep domain adaptation architecture where second-order statistics alignment technique is used to adapt features. DAN uses MMD to minimize the dissimilarity between source and target domains. DRCN introduces an unsupervised domain adaptation model which reconstruct source images that have a similar appearance to or qualities in common with the target images. RTN introduces residual transfer network where classifiers and features are adapted simultaneously. DAH uses deep hashing network for unsupervised domain adaptation. In DAH, MMD is utilized to decrease the dissimilarities between the source and target domains.

We use Caffe [12] framework to implement our proposed method. We use Alexnet architecture [15]. We conduct experiments with one NVIDIA GeForce GTX 1070 Graphics Processing Unit (GPU). For unsupervised domain adaptation techniques, we follow the standard protocol where the source data are labeled, but the target data are unlabeled. We make a comparison based on average classification accuracy for each transfer task.

As shown in Tables 1, 2 and 3, we compare the results of our proposed method with state-of-the-art approaches on three datasets (Office 31, Office-Home and Office-Caltech) in the context of classification accuracy. The classification accuracy of a model A_i depends on the images correctly identified. We evaluated all the methods using the following formula:

$$A_i = \frac{t}{n} \times 100, \tag{9}$$

where, t is the total number of correctly classified images, and n belong to the total images.

For Office-31 dataset, we report the image classification results in Table 1 for target data on different transfer tasks. In Table 2, the target data classification accuracy are reported for Office-Home dataset on twelve transfer tasks. For Office-Caltech dataset, the target classification accuracy on different transfer tasks are reported in Table 3. The accuracies stand for the percentage of correctly classified target images.

For Office-31 dataset, the previous best average result achieved by [6, 18] which are 73.7% and 73.6%, respectively. In contrast with their approach, our combined

Table 1 Image classification accuracies for deep domain adaptation on the Office-31 dataset. We use the standard protocol for unsupervised domain adaptation where source data are labeled, but target data are unlabeled. A-W indicates A (Amazon) is source and W (Webcam) is target

Methods	A-W	D-W	D-A	W-A	W-D	A-D	Avg.
TCA [23]	21.5	50.1	8.0	14.6	58.4	11.4	27.3
GFK [8]	19.7	49.7	7.9	15.8	63.1	10.6	27.8
VGG16 [29]	63.9	81.6	46.9	**54.1**	91.9	63.1	66.9
AlexNet [15]	53.4	79.9	46.9	47.5	84.1	55.6	61.2
DANN* [5]	**73.9**	94.9	–	–	99.5	–	-*
D-CORAL [32]	67.2	94.5	52.6	51.6	98.7	64.9	71.6
DAN [17]	68.5	96.0	50.0	49.8	**99.0**	66.8	71.7
DRCN [6]	68.7	96.4	**56.0**	54.9	99.0	66.8	73.6
RTN [18]	73.3	96.8	50.5	51.0	99.6	71.0	73.7
DAH [37]	68.3	96.1	55.5	53.0	98.8	66.5	73.0
Our method	72.1	**97.3**	54.6	53.9	98.7	**71.2**	**74.6**

DANN* reported three transfer tasks only

Table 2 Image classification accuracies for deep domain adaptation on the Office-Home dataset. We use the standard protocol for unsupervised domain adaptation where source data are labeled, but target data are unlabeled. Ar-Cl indicates Ar (Art) is source domain and Cl (Clipart) is target domain

Methods	A-C	A-P	A-R	C-A	C-P	C-R	P-A	P-C	P-R	R-A	R-C	R - P	Avg.
TCA [23]	19.93	32.08	35.71	19.00	31.36	31.74	21.92	23.64	42.12	30.74	27.15	48.68	30.34
GFK [8]	21.60	31.72	38.83	21.63	34.94	34.20	24.52	25.73	42.92	32.88	28.96	50.89	32.40
VGG16 [29]	30.40	**45.92**	**57.54**	35.40	48.67	50.75	**35.77**	30.51	60.20	49.62	34.54	64.00	45.28
AlexNet [15]	27.40	34.53	45.04	32.40	43.90	46.72	29.76	32.94	50.20	40.74	35.07	55.99	39.74
DANN [5]	33.33	42.96	54.42	32.26	49.13	49.76	30.49	**38.14**	56.76	44.71	42.66	64.65	44.94
D-CORAL [32]	32.18	40.47	54.45	31.47	45.8	47.29	30.03	32.33	55.27	44.73	42.75	59.40	42.79
DAN [17]	30.66	42.17	54.13	32.83	47.59	49.78	29.07	34.05	56.70	43.58	38.25	62.73	43.46
RTN [18]	31.23	40.19	54.56	32.46	46.60	48.25	28.20	32.89	56.38	45.53	44.74	61.28	43.53
DAH [37]	31.64	40.75	51.73	34.69	51.93	52.79	29.91	39.63	60.71	44.99	45.13	62.54	45.54
Our method	**35.15**	44.35	57.17	**36.82**	**52.45**	**53.67**	34.80	37.17	**62.15**	**49.95**	**46.29**	**66.05**	**48.00**

CORAL and MMD loss outperforms their results by 0.9% and 1.0% respectively. For Office-Home dataset, our proposed method achieves average 48.0% classification accuracy which outperforms most state-of-the-art approaches, such as, DAH [37] by 2.46%. For the Office-Caltech dataset, the existing best result was achieved by [18]. Our proposed method beats their average classification accuracy by 0.2%. Thus, the proposed model based on MMD and CORAL outperforms all comparison methods on most transfer tasks on the datasets. From Tables 1, 2 and 3, we can see that the proposed method achieves better average performance than other baseline conventional and deep domain adaptation methods.

Table 3 Image classification accuracies for deep domain adaptation on the Office-Caltech dataset. We use the protocol for unsupervised domain adaptation where source data are labeled, but target data are unlabeled. A-C indicates A (Amazon) is source and C (Caltech) is target

Methods	A-W	D-W	D-A	W-A	W-D	A-D	A-C	W-C	C-W	C-D	D-C	C-A	Avg.
TCA [23]	84.4	96.9	90.4	85.6	99.4	82.8	81.2	75.5	88.1	87.9	79.6	92.1	87.0
GFK [8]	89.5	97.0	89.8	88.5	98.1	86.0	76.2	77.1	78.0	77.1	77.9	90.7	85.5
AlexNet [15]	79.5	97.7	87.1	83.8	**100.0**	87.4	83.0	73.0	83.7	87.1	79.0	91.9	86.1
D-CORAL [32]	89.8	97.3	91.0	91.9	**100.0**	90.5	83.7	81.5	90.1	88.6	80.1	92.3	89.7
DAN [17]	91.8	98.5	90.0	92.1	**100.0**	91.7	84.1	81.2	90.3	89.3	80.3	92.0	90.1
RTN [18]	95.2	99.2	93.8	92.5	**100.0**	95.5	88.1	**86.6**	**96.9**	**94.2**	84.6	**93.7**	93.4
Our method	**95.7**	**99.4**	**94.7**	**94.8**	**100.0**	**96.6**	**89.1**	86.5	95.2	93.4	**84.7**	93.6	**93.6**

These results provide the suggestion that our proposed method is capable to acquire better classifiers which are adaptive in between domains and transferable features to solve domain adaptation issue.

From all the results in terms of image classification, we can find the following observations:

- Traditional deep learning approaches without domain adaptation perform better than the standard domain adaptation methods.
- The proposed unsupervised deep domain adaptation based on joint aligning of the second-order statistics and maximum mean discrepancy outperforms the state-of-the-art methods.
- Our models work better where the number of classes of objects is more. For example, Office-Home dataset contains 65 categories and we achieved 48% accuracy using our model.

4.4 Visualization

We use t-SNE for embedding visualization. To produce an embedding, we take images from Amazon and Webcam domains of Office-31 dataset. We use the CNN model to acquire the corresponding fc7-4096 dimensional vector for each image. After that, we plug these fc7-4096 vectors into t-SNE and generate 2-dimensional vector for each image. We plot a t-SNE embedding in Fig. 3 of images that are taken from Amazon and Webcam domains using our learned representation (right) and make a comparison it to an embedding formed with AlexNet in Fig. 3 (left). Examining the embeddings, we found that the clusters created by our model separate the classes while mixing the domains much more efficiently than the AlexNet approach where there is no domain adaptation technique is applied.

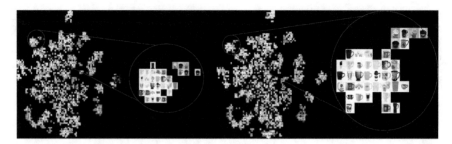

Fig. 3 t-SNE embedding of images that are taken from Amazon and Webcam domains using AlexNet model (left) and using two MMD and CORAL metrics in between fc7 and fc8 layers of the two stream CNN (right). While mixing domains, It is observed that the clusters created by our proposed model that can separate classes much more efficiently than AlexNet model where there is no domain adaptation technique is applied

5   Conclusion

In this chapter, we introduce an unsupervised deep domain adaptation architecture where the features and classifiers are adapted jointly. The source and target features are adapted by aligning covariances as well as maximum mean discrepancy and the classifiers are adapted by minimizing the entropy loss of the target data. Extensive Experimental results on standard benchmark datasets suggest the state-of-the-art performance. Prior deep domain adaptation techniques either use MMD or CORAL to decrease the mismatch between the source and target data. However, unlike previous work, we use both MMD and CORAL to adapt the features across domains. This makes our method a decent supplement to existing procedures.

References

1. Choi Y, Choi M, Kim M, Ha J-W, Kim S, Choo J (2018) Stargan: unified generative adversarial networks for multi-domain image-to-image translation. In: The IEEE conference on computer vision and pattern recognition (CVPR)
2. Dahl GE, Sainath TN, Hinton GE (2013) Improving deep neural networks for lvcsr using rectified linear units and dropout. In: International conference on acoustics, speech and signal processing (ICASSP)
3. Donahue J, Jia Y, Vinyals O, Hoffman J, Zhang N, Tzeng E, Darrell T (2014) Decaf: a deep convolutional activation feature for generic visual recognition
4. Fernando B, Habrard A, Sebban M, Tuytelaars T (2013) Unsupervised visual domain adaptation using subspace alignment. In: The IEEE conference on computer vision and pattern recognition (CVPR)
5. Ganin Y, Ustinova E, Ajakan H, Germain P, Larochelle H, Laviolette F, Marchand M, Lempitsky V (2016) Domain-adversarial training of neural networks. J Mach Learn Res 17(1)
6. Ghifary M, Kleijn WB, Zhang M, Balduzzi D, Li W (2016) Deep reconstruction-classification networks for unsupervised domain adaptation. Springer International Publishing, Cham

7. Girshick R, Donahue J, Darrell T, Malik J (2014) Rich feature hierarchies for accurate object detection and semantic segmentation. In: The IEEE conference on computer vision and pattern recognition (CVPR)
8. Gong B, Shi Y, Sha F, Grauman K (2012) Geodesic flow kernel for unsupervised domain adaptation. In: The IEEE conference on computer vision and pattern recognition (CVPR)
9. Goodfellow I, Bulatov Y, Ibarz J, Arnoud S, Shet V (2014) Multi-digit number recognition from street view imagery using deep convolutional neural networks
10. Hong W, Wang Z, Yang M, Yuan J (2018) Conditional generative adversarial network for structured domain adaptation. In: The IEEE conference on computer vision and pattern recognition (CVPR)
11. Hu L, Kan M, Shan S, Chen X (2018) Duplex generative adversarial network for unsupervised domain adaptation. In: The IEEE conference on computer vision and pattern recognition (CVPR)
12. Jia Y, Shelhamer E, Donahue J, Karayev S, Long J, Girshick R, Guadarrama S, Darrell T (2014) Caffe: convolutional architecture for fast feature embedding. In: ACM international conference on multimedia
13. Khatun A, Denman S, Sridharan S, Fookes C (2018) A deep four-stream siamese convolutional neural network with joint verification and identification loss for person re-detection. In: IEEE winter conference on applications of computer vision (WACV)
14. Koniusz P, Tas Y, Porikli F (2017) Domain adaptation by mixture of alignments of second-or higher-order scatter tensors. In: The IEEE conference on computer vision and pattern recognition (CVPR)
15. Krizhevsky A, Sutskever I, Hinton GE (2012) Imagenet classification with deep convolutional neural networks. In: Neural information processing systems (NIPS)
16. Lecun Y, Bengio Y, Hinton G (2015) Deep learning. Nature 521(7553):436–444, 5
17. Long M, Cao Y, Wang J, Jordan MI (2015) Learning transferable features with deep adaptation networks
18. Long M, Zhu H, Wang J, Jordan MI (2016) Unsupervised domain adaptation with residual transfer networks. In: Neural information processing systems (NIPS)
19. Long M, Zhu H, Wang J, Jordan MI (2017) Deep transfer learning with joint adaptation networks
20. Mancini M, Porzi L, Rota Bulò S, Caputo B, Ricci E (2018) Boosting domain adaptation by discovering latent domains. In: The IEEE conference on computer vision and pattern recognition (CVPR)
21. Murez Z, Kolouri S, Kriegman D, Ramamoorthi R, Kim K (2018) Image to image translation for domain adaptation. In: The IEEE conference on computer vision and pattern recognition (CVPR)
22. Pan SJ, Tsang IW, Kwok JT, Yang Q (2009) Domain adaptation via transfer component analysis. In: International joint conference on artificial intelligence (IJCAI)
23. Pan SJ, Tsang IW, Kwok JT, Yang Q (2011) Domain adaptation via transfer component analysis. IEEE Trans Neural Netw 22(2):199–210
24. Pan SJ, Yang Q (2010) A survey on transfer learning. IEEE Trans Knowl Data Eng 22(10):1345–1359
25. Patel VM, Gopalan R, Li R, Chellappa R (2015) Visual domain adaptation: a survey of recent advances. IEEE Signal Process Mag 32(3)
26. Pinheiro PO (2018) Unsupervised domain adaptation with similarity learning. In: The IEEE conference on computer vision and pattern recognition (CVPR)
27. Saenko K, Kulis B, Fritz M, Darrell T (2010) Adapting visual category models to new domains. In: European conference on computer vision (ECCV)
28. Saito K, Watanabe K, Ushiku Y, Harada T (2018) Maximum classifier discrepancy for unsupervised domain adaptation. In: The IEEE conference on computer vision and pattern recognition (CVPR)
29. Simonyan K, Zisserman A (2014) Very deep convolutional networks for large-scale image recognition. CoRR, abs/ arXiv:1409.1556

30. Sun B, Feng J, Saenko K (2016) Return of frustratingly easy domain adaptation. In: AAAI conference on artificial intelligence
31. Sun B, Feng J, Saenko K (2017) Correlation alignment for unsupervised domain adaptation. In: Domain adaptation in computer vision applications, pp 153–171
32. Sun B, Saenko K (2016) Deep coral: correlation alignment for deep domain adaptation. In: European conference on computer vision workshops
33. Sutskever I, Vinyals O, Le QV (2014) Sequence to sequence learning with neural networks. In: Neural information processing systems (NIPS)
34. Torralba A, Efros AA (2011) Unbiased look at dataset bias. In: The IEEE conference on computer vision and pattern recognition (CVPR)
35. Tzeng E, Hoffman J, Darrell T, Saenko K (2015) Simultaneous deep transfer across domains and tasks
36. Tzeng E, Hoffman J, Zhang N, Saenko K, Darrell. T (2014) Deep domain confusion: maximizing for domain invariance. *CoRR*, abs/ arXiv:1412.3474
37. Venkateswara H, Eusebio J, Chakraborty S, Panchanathan S (2017) Deep hashing network for unsupervised domain adaptation. In: The IEEE conference on computer vision and pattern recognition (CVPR)
38. Volpi R, Morerio P, Savarese S, Murino V (2018) Adversarial feature augmentation for unsupervised domain adaptation. In: The IEEE conference on computer vision and pattern recognition (CVPR)
39. Yan H, Ding Y, Li P, Wang Q, Xu Y, Zuo W (2017) Mind the class weight bias: weighted maximum mean discrepancy for unsupervised domain adaptation. In: The IEEE conference on computer vision and pattern recognition (CVPR)
40. Yang B, Ma AJ, Yuen PC (2018) Learning domain-shared group-sparse representation for unsupervised domain adaptation. Pattern Recognit 81:615–632

Multi-modal Conditional Feature Enhancement for Facial Action Unit Recognition

Nagashri N. Lakshminarayana, Deen Dayal Mohan, Nishant Sankaran, Srirangaraj Setlur and Venu Govindaraju

Abstract Current state-of-the-art methods in multi-modal fusion typically rely on generating a new shared representation space onto which multi-modal features are mapped for the goal of obtaining performance improvements by combining the individual modalities. Often, these heavily fine-tuned feature representations would have strong feature discriminability in their own spaces which may not be present in the fused subspace owing to the compression of information arising from multiple sources. To address this, we propose a new approach to fusion by enhancing the individual feature spaces through information exchange between the modalities. Essentially, domain adaptation is learnt by building a shared representation used for mutually enhancing each domain's knowledge. In particular, the learning objective is modeled to modify the features with the overarching goal of improving the combined system performance. We apply our fusion method to the task of facial action unit (AU) recognition by learning to enhance the thermal and visible feature representations. We compare our approach to other recent fusion schemes and demonstrate its effectiveness on the MMSE dataset by outperforming previous techniques.

Keywords Feature fusion · Feature fine-tuning · Facial action unit recognition · Deep fusion · Multi-modal representation learning

N. N. Lakshminarayana, D. D. Mohan, N. Sankaran—Equal contribution authors listed in alphabetical order.

N. N. Lakshminarayana · D. D. Mohan (✉) · N. Sankaran · S. Setlur · V. Govindaraju
University at Buffalo, Buffalo, NY 14226, USA
e-mail: dmohan@buffalo.edu

N. N. Lakshminarayana
e-mail: nagashri@buffalo.edu

N. Sankaran
e-mail: n6@buffalo.edu

S. Setlur
e-mail: setlur@buffalo.edu

V. Govindaraju
e-mail: govind@buffalo.edu

© Springer Nature Switzerland AG 2020
R. Singh et al. (eds.), *Domain Adaptation for Visual Understanding*,
https://doi.org/10.1007/978-3-030-30671-7_7

1 Introduction

In recent years, methods that combine features from multiple data sources have been gaining popularity. This can be primarily attributed to the large amount of data produced by different multi-modal sensors. Most of the methods that try to combine the multi-modal data, broadly fall under two major categories. Multi-view learning methods [25], which look at finding a subspace or a shared space between the data of multiple modalities and employ that as a unified representation. These methods generally try to enforce a constraint that increases the similarity of features learned by the views. On the other hand, multi-modal fusion methods [14], try to combine features of different representations to improve performance of the overall system.

Majority of the multi-modal fusion methods try to create a unified feature space. This is done either by mapping the current feature space to a higher dimension or by learning a latent representation after concatenating multi-modal features. Due to the recent advancements in new embedding methods such as [20] and complex multi dataset training procedures for deep learning, the features produced by such networks are highly optimal. Generating a unified representation from multiple such features might require equally complex training procedures. In this chapter, we try to rethink the premise of the necessity of having a unified representation. We theoretically and experimentally show that learning linear transformation that increase separability of these features in their respective feature spaces can be an alternative to existing methods. In order to validate our claims, we apply our method to the problem of recognition of facial action units.

Facial expressions are one of the most important nonverbal cues in any inter-personal communications. Facial expressions can be measured in two dimensions popularly. The judgmental coding system describes emotions in a latent emotion space. The frequently used parameters in this scheme are the seven universal emotions, namely, Anger, Fear, Disgust, Happiness, Sadness, Surprise, and Contempt. A more elaborate way of describing emotions is using the FACS coding scheme. In their chapter, [5] define FACS as a measure of different facial muscle movements that contribute to the facial expressions either independently or in pairs. Each of the action Units describes a movement or contraction of the facial muscle. Thus, they encode the anatomically visible changes rather than relying on the observer's inference of emotions. The granularity of FACS is particularly beneficial in detecting even the subtle controlled or uncontrolled facial behavior. For several decades, they have been extensively used in forensics, neuro-marketing, health care, etc. Although facial action units' recognition is well explored in the visible light domain (VLD), the RGB images suffer from illumination changes and can only capture the visual changes that occur as an effect of the AUs. There are, however, some physiological changes that the face undergoes during the occurrence of the action units, such as skin temperature changes, variation in the heart rate, blood pressure, and respira-tion rate that can not be captured using the visible images. In contrast, the infrared images that allow detection of skin temperature variations and are invariant to illu-mination changes and skin tone variation from person to person, have been shown to be sensitive to AU movements [13]. Thus, the visible and thermal images encode complementary aspects of facial action units.

In this chapter, we propose an alternative to the current existing multi-modal fusion methods. We train a DenseNet Model on both the visible and thermal images and generate corresponding visible and thermal features. We present an idea of enhancing existing feature spaces by only applying scaling and translation perturbation. The perturbation that is to be applied to each feature is learned by the network by jointly looking at all the feature representations. By doing so, we generate an enhanced feature representation of the original thermal and visible features. These enhanced features when combined, improve the overall performance of the system.

2 Related Work

There is a large amount of literature on the design and application of techniques used for combining multiple features to improve the overall performance of any system. The full treatment of these fusion techniques is well beyond the scope of this chapter. However, we will look at some of the recent methods that have focused on fusing multi-modal data to provide the proper context to our work. Simplistic models that use linear combinations of features are insufficient to capture the complex correlations between the modalities. One such approach is aggregating features weighted by the metadata [18] pertaining to face images, for the task of face recognition. Another approach is to use bilinear pooling [16] which captures the second-order statistics of the features. High dimensionality, vast parameterization, and slow convergence limit the practical applicability of these algorithms. To address these issues, [26] proposed a multi-modal factorized bilinear pooling (MFB) approach used for a Visual Question-answering task. Although it provides compact features and robust performance, the method attempts to learn a completely new feature space. Recently, in [1], the authors use a deep fusion network to jointly represent heterogeneous features from face images by performing nonlinear transformations of the concatenated feature space. Zhao et al. [29] present, as part of a CNN system for person reidentification, a tree structure-based fusion network that encourages competition among features arising from decomposed image regions (such as legs, arms, head, etc.) during fusion to select relevant features. They implement this strategy by performing element-wise max pooling operations on feature inputs and transforming the resulting feature activations to representations utilized by higher stages of the fusion structure. For solving the problem of missing features and curse of dimensionality in multi-modal fusion, the authors of [11] propose to fuse the multi-modal features by grouping them into a set of subspaces represented as a point on a Grassmann manifold and employing the L2 Hausdorff distance for comparing feature vectors with different number of subspaces. Heterogeneous feature structure fusion [15] jointly optimizes the internal (within each feature set) and external structures (across different feature sets) explicitly via a unified feature projection. Specifically, the algorithm represents the internal structure using a locality preserving projection (LPP) and the external structure by canonical correlation analysis (CCA) and is optimized via linear programming or eigenvector methods. The proposed multi-modal conditional feature enhancement method (MCFE) differs from the previous literature in that, rather than learning a new high dimensional feature space, MCFE uses a feature enhancement technique to improve the discriminative capabilities of each of the representations using minimal perturbations.

Facial expressions can be measured in two dimensions popularly. The Judgmental Coding System describes expressions in a latent emotion space such as anger, fear, disgust, happiness, etc. A more elaborate way of describing emotions is using the FACS coding scheme. They encode the anatomically visible changes rather than relying on the observer's inference of emotions. Recently the availability of spontaneous expression datasets has made it possible for researchers to implement automatic facial action unit recognition techniques. A detailed survey of the different methods and their performance is listed in [3, 24]. Owing to the popularity of deep learning in visual tasks, currently, most of the work in facial action unit recognition in the visible light domain, learn CNN-based features. In this chapter, we review some of the recent literature in facial action unit recognition. In their chapter, [7] use a seven-layer network to detect the occurrence and intensity of facial action units. In [6], authors use multi-label CNNs to extract appearance based features. Following the intuition that certain facial regions are more import than others for a particular AU, patch learning methods [30] have been used popularly. Sizhong et al. [8] introduced an incremental boosting layer on top of a three-layer CNN to deal with limited positive samples in each class. The introduced IB-CNN is trained separately for each class and does not take into account the interdependencies and correlations of different AUs. Action units can also be analyzed by exploring temporal domain along with the spatial. The authors of [12] used a combination of CNN and bidirectional LSTM to jointly learn the shape, appearance, and dynamic features. The CNN had two input streams for sequence of image regions and sequence of corresponding binary masks merged after the first pooling layer followed by two convolution layers. In [2] use two layers of LSTM on top of CNN to extract spatiotemporal information. Most of the aforementioned methods use common architectures of CNNs like AlexNets, VGG nets [21] or their simple modifications. In this chapter, we use the recent Densenet-121 architecture for feature extraction, with a few modifications like multitask learning and a weighted cross-entropy loss. The existing work on facial action unit recognition extract features from unimodal data. However, data from different modalities could contain complementary information and learning their correlations could enrich the feature space. In 2011, [13] analyzed the thermal fluctuations subjective to different AUs. Through their study, they found thermography to be a promising alternative discriminator for AUs. However, the lack of any publicly available dataset with synchronous visible and thermal videos with AU annotations impaired the thermal analysis of AUs. In 2016, [28] presented a multi-modal spontaneous emotion (MMSE) dataset consisting of 2D, 3D videos, thermal videos, and other physiological signals like heart rate, electrical conductivity of skin, etc. The dataset consists of 140 subjects and for each participant, 2D videos, 3D videos, and thermal videos from IR sensors are collected for ten tasks designed to elicit spontaneous emotions. In this chapter, we use the MMSE dataset to learn a multi-modal conditional feature enhancement representation for thermal and visible images.

3 Approach

3.1 Overview

The schematic of the proposed multi-modal conditional feature enhancement (MCFE) method is shown in Fig. 1. MCFE consists of a feature extraction stage followed by a feature enhancement stage. Addressing the task of facial action unit recognition, we first train a deep CNN that learns to assign action unit labels for a given RGB frame focused on an individual's face. This network is optimized using a multitask learning framework with class weighting incorporated, to solve the issue of class imbalance prevalent in such problems. While one network is trained on the visible spectrum, another network is trained similarly on the thermal spectrum for the same task. Each network learns a specific view of the task and we implement a novel multi-modal learning solution to *enhance* their corresponding representations by modeling their correlations in the shared subspace. Contrary to traditional fusion approaches, our approach does not attempt to create a unified fused representation of the modalities that is better equipped at solving the task. Rather, in our fusion approach, we emphasize transferring information that is uniquely learnt from the individual modalities to other view representations with the aim of improving the performance of each modality's feature representation oriented toward maximizing the combined system's performance. This method has an advantage over traditional

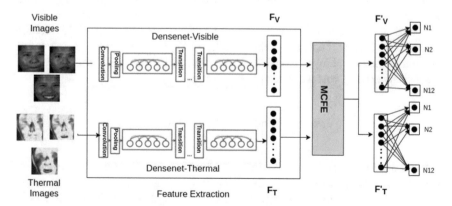

Fig. 1 Overview of the proposed MCFE framework. The system extracts Deep features (F_v and F_T) from paired visible and thermal images simultaneously using their corresponding DenseNet models. Further, using the proposed approach, the enhanced features (F'_v and F'_T) are produced

fusion schemes in that, instead of vastly increasing the search space for finding an optimum representation that describes both modalities equally, it limits the problem to only determining the corrections/perturbations it needs to apply to each view's representation guided by the accompanying views. In doing so, we enhance not only the individual representations but also the overall system performance which aggregates the performances of the individual modalities. In the following sections, we detail the approaches that we used for extracting action unit features and our novel fusion approach applied to multi-modal facial action unit recognition.

3.2   Feature Extraction

Research in facial expression recognition can be categorized mainly on the basis of feature extractors and classifiers. The ability of handcrafted features like SIFT [17], HOG [4] etc to capture the complex nonlinear transformations of the face caused by expressions are limited. CNNs on the other hand, has shown the ability to learn optimal features for vision-based tasks like handwritten character recognition, face recognition, etc. A series of convolutional filters can extract features starting from abstract information like edges to complex patterns like faces in the subsequent layers. However, with deeper structures, the gradient vanishes as it reaches the beginning layers. Networks with short skip connections like ResNets [9] and Highway Networks [22] prevent this by providing an alternative and easier way for gradients to flow. Following this, DenseNet [10] was introduced wherein features from one layer are connected to features of all the preceding layers. As a result, the lower level abstract features are combined with the higher level granular features. Although DenseNets learn a representation similar to the deeper models, owing to its compact parameterization, it is less prone to over fitting, and enables feature reuse. To this end, we use the Densenet-121 architecture with modifications for extracting features from both visible and thermal images. The network consists of four dense blocks followed by an output layer consisting of 12 neurons for 12 classes. The sigmoid activation is used at the final classification layer. Typically the cross entropy loss is applied to the output layer. Consider N AU classes, then for each input, the multi-class cross-entropy cost is calculated as follows:

$$C = \sum_{i=1}^{N} (y_i' \log y_i + (1 - y_i') \log(1 - y_i)) \tag{1}$$

In the above formulation, the individual components of loss corresponding to each AU is given equal weight. Most of the facial action unit datasets are heavily imbalanced. Some of the action units have very low positive to negative sample ratio, otherwise called the occurrence rate. Therefore, in order to account for the under represented classes, the individual loss components needs to be weighted. However, calculating the weights for each class with respect to other classes for a multi-label classification problem can be quite complex as each data sample can contain more than one AU class. Therefore to overcome this problem, we use a multitask framework wherein a separate binary cross-entropy loss is applied to each of the N output neurons and the weights applied to the loss components are weighted by the ratio of their respective positive and negative samples. The final output layer of the DenseNet is split into 12 output neurons. The cost function is calculated as follows:

$$b_i = w_i * (y_i' \log y_i) + 1 * ((1 - y_i')log(1 - y_i)) \qquad (2)$$

The negative samples are weighted by 1 and the positive samples are weighted by w_i given as

$$w_i = \frac{n_i}{p_i} \qquad (3)$$

where n_i is the total negative samples for AU_i and p_i is the total positive samples. The final loss is calculated as the sum individual binary cross-entropy losses:

$$C_i = \sum_{i=1}^{N} b_i \qquad (4)$$

Thus, the loss formulation takes into account the individual class distributions, while still learning the correlations between the different action units. The network is trained with random weight initialization instead of initializing with the typical ImageNet classification weights.

3.3   Multi-modal Conditional Feature Enhancement (MCFE)

Deep multi-modal fusion has typically relied on learning the feature correlations among the modalities by stacking a number of fully connected layers applied on a merged representation (concatenation, sum, etc.) or by projecting each modality's feature space onto a common optimal subspace for the specific task. Such methods eventually arrive at a new shared representation for fusion, but is it necessary to have to construct a completely new representation? We address this question by proposing to employ the existing feature space and design a fusion scheme that, based upon a shared representation, learns to only modify or perturb the original features in such a way as to improve feature separability in their existing feature spaces.

Consider k input modalities $x_i, i = 1, .., k$ and their corresponding feature representations are obtained as follows:

$$v_i = f(x_i; \theta_i) \qquad (5)$$

where $v_i \in \mathbb{R}^{d_i}$, f may be an MLP, DNN or other feature extractors and θ_i are the parameters for the corresponding modalities which may be shared. We define a function g with parameters ∇ which transforms all the input modalities' features into a latent representation $l \in \mathbb{R}^n$ thus:

$$l = g(v_1, v_2, ...v_k; \nabla) \tag{6}$$

Based on this latent representation, we compute M transformation factors (feature wise scaling and translation) $s_i = [s_i^1, .., s_i^M]$ and $t_i = [t_i^1, .., t_i^M]$ for each modality i as below (omitting the subscript i for brevity)

$$s^j = \sigma(W_s^{j^T} l + b_s^j) \tag{7}$$

$$t^j = \sigma(W_t^{j^T} l + b_t^j) \tag{8}$$

Since the above equations are for each modality, there are k weights and biases W_s^j and b_s^j corresponding to the scaling factors and k weights and biases W_t^j and b_t^j corresponding to the translation factors with $j = 1, .., M$ and σ denoting the sigmoid nonlinearity. With these, we can construct M different variants of each feature vector v_i as

$$e_i^j = (s_i^j \odot v_i) \oplus t_i^j \tag{9}$$

Finally we choose 1 out of the M different enhanced features e_i^j by predicting *importance weights* ch_i^j for the M variants and running it through a softmax activation to pick the most relevant enhanced feature vector e_i^*:

$$ch_i^j = softmax(W_c^{j^T} l + b_c^j) \tag{10}$$

$$e_i^* = \sum_{j=1}^{M} ch_i^j * e_i^j \tag{11}$$

where W_c^j and b_c^j are the k weights and biases corresponding to the choice prediction function applied to the k modalities. This final enhanced feature representation is presented to the classification layer for improved performance on the task being solved. Figure 2 illustrates the general architecture of the proposed multi-modal conditional feature enhancement system.

The presence of M variants of transformation factors and consequently M different versions of the enhanced features enables the learning algorithm to explore a variety of improvements that can be applied to the original feature set. This is similar in concept to the multiple paths that are present in the inception [23] deep network architecture, where each layer gets its data from several different *views* of the previous layer's output. The presence of such multiple paths of computation directly improves gradient flow to the prior layers since each path would begin to learn aspects unique to a particular view for better representing the solution space. Additionally, the ability to select the variant that is most suitable for the given sam-

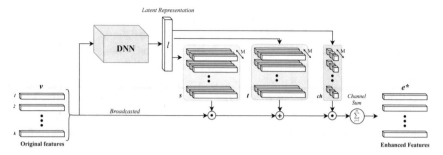

Fig. 2 MCFE Architecture. The system takes k modality features v and proposes M variants of element-wise scaling s and translation t parameters applied on the input features. It finally weights the aggregation of the proposals using ch via the channel sum operation to finally arrive at k enhanced features $e*$

ple of multi-modal features enables conditioning the fusion methodology applied to the features. This allows the gradients to pass through multiple parallel routes to the latent representation mapping function g and thus allows faster convergence as observed in the experiments. It should be noted that the latent representation l being learnt is different from the traditional shared representations that prior works obtain during fusion. This is because in this formulation, l encapsulates information relevant only to making a decision as to how to *correct* every modality's feature representation independently rather than representing a common unified subspace which is used directly for the classification task. There are benefits for defining the latent representation in such a manner and this will be explored in Sect. 3.4.

3.4   Training MCFE for AU Recognition

MCFE is applied to the thermal and visible features arising out of the corresponding thermal and RGB frames extracted as described in Sect. 3.2. The latent mapping function g that operates on the two inputs is implemented as a modified lightweight DenseNet CNN. The latent representation l is connected to three separate fully connected layers—one each for scaling, translation, and predicting choice. Upon computing the final enhanced feature set e_i* as described in Sect. 3.3 they are independently connected to separate AU classification layers (one each for thermal and visible spectrum) composed of 12 outputs each indicating the presence of a specific action unit for the given image. In order to orient the training of MCFE's parameters to the final goal of improving the overall system (thermal and visible combined) performance and not just the individual modality's performance, we perform a simple average of the predictions by the thermal and visible classification layers. By combining the predictions and optimizing against them, the network is encouraged to learn parameters that would enable a particular modality's enhanced feature to compensate for the shortcomings of the other feature.

In the naive approach, we can initialize the network with random weights. However, this would either fail to converge or provide unsatisfactory results. Intelligent initializations help the network perform significantly better than traditional approaches. We take two steps in this direction: (a) initialize the scaling layer to produce an output of 1s and the translation layer to produce an output of 0s; and (b) initialize the classification layers with the corresponding pretrained weights learnt during the feature extraction step. Performing Step (a) ensures that in the initial state-of-the network the enhanced features are the same as the original features, forming an identity map:

$$e_i^j = (s_i^j \odot v_i) \oplus t_i^j$$
$$= (1 \odot v_i) \oplus 0 = v_i$$

which gives a good starting location to begin gradient descent. Since the overall error reduces in a specific direction, there can only be an improvement to the feature representation (or at worst can remain the same). Step (b) accomplishes the objective of reusing the information learnt from prior training to give a good target representation that the previous layers can attempt to produce from the original representation. One question that arises is how can we ensure that the enhanced features are indeed only *corrected* versions of the original features and not entirely a new set of features learnt by the network? We address this by freezing the classification layer weights from training (which are already preinitialized). This enforces a constraint on the input to the classification layers (i.e., the enhanced features) such that it adheres to the general feature space representation that the classification layer has been trained to operate on. Imposing this constraint limits the optimizer's task to finding δs to each individual feature so as to magnify or diminish their effect and in doing so make the feature space more separable with respect to the given classification layer parameters. This is a much simpler task to optimize for, than having to explore unconstrained combinations of feature spaces and, as we observe in the experiments, leads to much faster convergence with strong performance gains. Additionally, with the formulation of a single latent representation that is responsible for the computation of three separate quantities, it implements the information bottleneck principle which has been shown to produce better generalizations [19] and also mirrors the multitask learning framework's principle of leveraging a single representation for accomplishing multiple disparate tasks.

4   Experiments

4.1   Datasets

We used the Multi-modal Spontaneous Emotion (MMSE) database to evaluate our performance. MMSE contains 2D and thermal videos of 140 participants from ten tasks, each eliciting different emotions. Among them, only four tasks were labeled for facial action units. Expert AU coders annotated each frame using the Facial Action Coding System. The thermal sensor and the RGB camera were mounted on top of each other and their frame rates were set to 25 fps for synchronization. In our experiments, we used all 196, 793 visible frames, and 195, 411 thermal frames. Out of the available images, only 133, 309 paired frames were available for our multi-modal learning experiments.

4.2 Settings

Preprocessing All the input images in the dataset were aligned using the MTCNN framework [27] based on the 49 facial landmark points provided by the MMSE dataset. Further, the images were cropped to 170×170 and randomly rotated for data augmentation. The presence of each action unit was labeled as $+1/0$. The data samples with missing labels and faces were excluded from training.

Network Settings and Training for Feature Extraction We adopted a threefold cross validation protocol to train our networks. For each experiment, we split the dataset into three subject dependent partitions using two partitions for training and the remaining one partition for validation. Both CNNs in our implementation are trained using the weighted cross-entropy loss defined in Eqs. 2–4 The models are trained with SGD as the optimizer with learning rate initialized to $1e - 3$.

Network Settings and Training for MCFE We use DenseNet-100x12 architecture, which has a depth of 100 and a growth rate of 12 for getting the latent representation from the features. Along with the DenseNet, the network consists of three fully connected layers corresponding to scaling, translation and prediction, and two classification layers corresponding to the two modalities, resulting in a total of 24M parameters. The network is trained using Adam optimizer, with a learning rate of 0.1 and a batch size of 128 for 32 epochs.

4.3 Results

We show the results of our experiment in Table 1. We report our performance using the F1 metrics widely used in the literature of facial action unit recognition. For each experiment, we report the individual F1 scores of each action unit and also the average F1 score across all action units. We report the average F1 scores across three splits. We compare MCFE performance with other relevant state-of-the-art multi model fusion strategies such as MFB [26] and Concatenation network [1]. We also compare our enhanced features with the original features from a simple sum score level fusion perspective.

From Table 1, we note that the fusion results of the proposed MCFE methods outperform the other methods. The simple score fusion of the enhanced features

Table 1 Threefold cross validation results on the MMSE dataset over 12 action units. *Visible* and *Thermal* refer to the results obtained using the feature extraction method described in Sect. 3.2. and *Average* refers to the performance obtained by averaging their predictions. *ConcatFusion* is MLP-based fusion on the concatenated thermal and visible features as defined in [1], *MFB* is multi-modal factorized binary pooling described in [26]. *MCFE Visible* and *MCFE Thermal* are the enhanced features for the corresponding modalities produced by MCFE while *MCFE Average* refers to the performance obtained by averaging the enhanced features' predictions

AU	MMSE threefold validation results														Mean F1
	1	2	4	6	7	10	12	14	15	17	23	24			
Visible	0.588	0.572	0.488	0.902	0.933	0.945	0.931	0.906	0.555	0.579	0.643	0.504			0.712
Thermal	0.397	0.362	0.225	0.865	0.873	0.919	0.895	0.845	0.446	0.326	0.457	0.179			0.565
Average	0.623	0.592	0.489	0.909	0.936	0.950	0.934	0.913	0.578	0.565	0.642	0.465			0.711
ConcatFusion [1]	0.617	0.573	0.274	0.914	0.934	0.944	0.929	0.905	0.562	0.563	0.625	0.242			0.673
MFB [26]	**0.629**	**0.605**	0.516	**0.917**	0.937	0.949	0.930	0.904	0.556	0.579	0.637	**0.486**			0.720
MCFE visible	0.593	0.576	0.485	0.903	0.935	0.946	0.931	0.910	0.554	0.580	0.645	0.512			0.714
MCFE thermal	0.411	0.379	0.283	0.879	0.899	0.921	0.895	0.853	0.445	0.406	0.532	0.189			0.591
MCFE average	0.626	0.602	**0.531**	0.912	**0.938**	**0.950**	**0.933**	**0.907**	**0.577**	**0.597**	**0.646**	0.484			**0.725**

Table 2 Variation of the mean F1 score of split 2 of the data set for different values of M. M refers to the number of scaling and translation values initially generated from network, from which the choice layer picks the best value

M mean	F1 score
2	0.7714
4	0.7688
8	0.7711
16	0.7662
32	0.7704

provides a performance improvement of about 1.1% on average, compared to the original features. However, we also note that even though MFB performs slightly better on some action units, MCFE significantly outperforms MFB in the action units which are severely underrepresented in the dataset. It is appealing to note that this improvement is obtained without using any explicit class balancing strategies. Another interesting observation is the significant performance improvement of thermal features across all the action units. This can be attributed to the fact that the network would have learned good translation and scaling factors for the thermal features so as to align it better with the visible component. Even though there is a slight reduction in the performance of visible features in some cases, this can be seen as a trade-off, as the system tries to align features from both the modalities, so as to improve the overall performance.

Additionally, we perform experiments with varying values of M and the results are provided in Table 2. Since there is no clear correlation between the value of M and the performance of the system that can be observed, M can be treated as any other hyperparameter which depends on the input data, different parameters of the network, and the task being optimized for.

5   Future Work

In this work, we explored how to maximally separate multi-modal features in their respective feature space. We used a simple sum-based score fusion as a method to improve the overall performance of the system. In the current work, we did not explore the amount of complementary information present in the multi-modal features. It will be interesting to investigate a fusion methodology which tries to enhance individual features by increasing the complementarity between them. We created this fusion methodology for a multi-modal feature scenario. However, we believe this fusion methodology could be adapted to enhance the performance of systems, which are based on an ensemble of classifiers. Furthermore, MCFE is performed on spatial features treating individual frames as a data sample. In future, we aim to apply our fusion methodology to features modeling both spatial and temporal aspects in a multi-label setting. The proposed MCFE is a general fusion technique and can be applied to other tasks that can make use of the multi-modal data.

6 Conclusion

Multi-modal fusion techniques are being used extensively in combining information from multiple data sources. Several approaches have been proposed, which try different techniques to create a unified representation directly for fusion. However, the underlying concept of requiring a final unified representation has remained unchanged. In this chapter, we proposed an alternative to the concept of a joint unified representation. Through theoretical and experimental validation, we find that we can learn the factors that help to better align the features in their respective feature spaces to maximize separability. In doing so, we eliminate the need for a unified representation. We also show that such aligned features can be easily combined so as to improve the overall performance of the system.

Acknowledgements This material is based upon work partially supported by the National Science Foundation under Grant IIP #1266183.

References

1. Bodla N, Zheng J, Xu H, Chen J, Castillo CD, Chellappa R (2017) Deep heterogeneous feature fusion for template-based face recognition. CoRR http://arxiv.org/abs/1702.04471
2. Chu WS, De la Torre F, Cohn JF (2017) Learning spatial and temporal cues for multi-label facial action unit detection. In: 2017 12th IEEE international conference on automatic face and gesture recognition (FG 2017). IEEE, pp 25–32
3. Corneanu CA, Simón MO, Cohn JF, Guerrero SE (2016) Survey on rgb, 3D, thermal, and multimodal approaches for facial expression recognition: history, trends, and affect-related applications. IEEE Trans Pattern Anal Mach Intell 38(8):1548–1568
4. Dalal N, Triggs B (2005) Histograms of oriented gradients for human detection. In: 2005 IEEE computer society conference on computer vision and pattern recognition (CVPR'05), vol 1, pp 886–893. https://doi.org/10.1109/CVPR.2005.177
5. Ekman P, Friesen WV (1976) Measuring facial movement. Environ Psychol Nonverbal Behav 1(1):56–75
6. Ghosh S, Laksana E, Scherer S, Morency LP (2015) A Multi-label convolutional neural network approach to cross-domain action unit detection. In: Proceedings of ACII 2015. IEEE, Xi'an, China. http://ict.usc.edu/pubs/A%20Multi-label%20Convolutional%20Neural%20Network%20Approach%20to%20Cross-Domain%20Action%20Unit%20Detection.pdf
7. Gudi A, Tasli HE, Den Uyl TM, Maroulis A (2015) Deep learning based facs action unit occurrence and intensity estimation. In: Proceedings of the 2015 11th IEEE international conference and workshops on automatic face and gesture recognition (FG), vol 6. IEEE, pp 1–5
8. Han S, Meng Z, Khan AS, Tong Y (2016) Incremental boosting convolutional neural network for facial action unit recognition. In: Advances in neural information processing systems, pp 109–117

9. He K, Zhang X, Ren S, Sun J (2016) Deep residual learning for image recognition. In: Proceedings of the IEEE conference on computer vision and pattern recognition, pp 770–778

10. Huang G, Liu Z, Weinberger KQ, van der Maaten L (2017) Densely connected convolutional networks. In: Proceedings of the IEEE conference on computer vision and pattern recognition, vol 1, p 3

11. Huang H, Liu H, Kong X, Lou X, Wang Z (2017) Heterogeneous massive feature fusion on grassmannian manifold. J Phys: Conf Ser 887:012066. (IOP Publishing)

12. Jaiswal S, Valstar M (2016) Deep learning the dynamic appearance and shape of facial action units. In: 2016 IEEE winter conference on applications of computer vision (WACV). IEEE, pp 1–8

13. Jarlier S, Grandjean D, Delplanque S, N'diaye K, Cayeux I, Velazco MI, Sander D, Vuilleumier P, Scherer KR (2011) Thermal analysis of facial muscles contractions. IEEE Trans Affect Comput 2(1):2–9

14. Lahat D, Adalı T, Jutten C (2015) Multimodal data fusion: an overview of methods, challenges and prospects. Proc IEEE 103(9):1449–1477. https://hal.archives-ouvertes.fr/hal-01179853

15. Lin G, Fan G, Kang X, Zhang E, Yu L (2016) Heterogeneous feature structure fusion for classification. Pattern Recognit. 53:1–11

16. Lin TY, RoyChowdhury A, Maji S (2015) Bilinear CNN models for fine-grained visual recognition

17. Lowe DG (2004) Distinctive image features from scale-invariant keypoints. Int J Comput Vis 60(2):91–110. https://doi.org/10.1023/B:VISI.0000029664.99615.94

18. Sankaran N, Tulyakov S, Setlur S, Govindaraju V (2018) Metadata-based feature aggregation network for face recognition. In: 2018 11th IAPR international conference on biometrics (ICB 2018). IEEE

19. Saxe AM, Bansal Y, Dapello J, Advani M, Kolchinsky A, Tracey BD, Cox DD (2018) On the information bottleneck theory of deep learning. In: International conference on learning representations

20. Schroff F, Kalenichenko D, Philbin J (2015) Facenet: a unified embedding for face recognition and clustering. In: Proceedings of the IEEE conference on computer vision and pattern recognition, pp 815–823

21. Simonyan K, Zisserman A (2014) Very deep convolutional networks for large-scale image recognition. arXiv preprint arXiv:1409.1556

22. Srivastava RK, Greff K, Schmidhuber J (2015) Training very deep networks. In: Advances in neural information processing systems. pp 2377–2385

23. Szegedy C, Ioffe S, Vanhoucke V, Alemi AA (2017) Inception-v4, inception-resnet and the impact of residual connections on learning

24. Tian YL, Kanade T, Cohn JF (2005) Facial expression analysis. In: Handbook of face recognition. Springer, Berlin, pp 247–275

25. Xu C, Tao D, Xu C (2013) A survey on multi-view learning. arXiv preprint arXiv:1304.5634

26. Yu Z, Yu J, Fan J, Tao D (2017) Multi-modal factorized bilinear pooling with co-attention learning for visual question answering. CoRR http://arxiv.org/abs/1708.01471

27. Zhang K, Zhang Z, Li Z, Qiao Y (2016) Joint face detection and alignment using multitask cascaded convolutional networks. IEEE Signal Process Lett 23(10):1499–1503

28. Zhang Z, Girard JM, Wu Y, Zhang X, Liu P, Ciftci U, Canavan S, Reale M, Horowitz A, Yang H, Cohn JF, Ji Q, Yin L (2016) Multimodal spontaneous emotion corpus for human behavior analysis. In: 2016 IEEE CVPR, pp 3438–3446. https://doi.org/10.1109/CVPR.2016.374

29. Zhao H, Tian M, Sun S, Shao J, Yan J, Yi S, Wang X, Tang X (2017) Spindle net: person re-identification with human body region guided feature decomposition and fusion. In: Proceedings of the IEEE conference on computer vision and pattern recognition. pp 1077–1085

30. Zhao K, Chu WS, De la Torre F, Cohn JF, Zhang H (2016) Joint patch and multi-label learning for facial action unit and holistic expression recognition. IEEE Trans Image Process 25(8):3931–3946

Intuition Learning

Anush Sankaran, Mayank Vatsa and Richa Singh

Abstract "By reading only the title and abstract, do you think this research will be accepted in an AI conference?" A common impromptu reply would be "I don't *know* but I have an *intuition* that this research might get accepted". Intuition is often employed by humans to solve challenging problems without explicit efforts. Intuition is not trained but is learned from one's own experience and observation. The aim of this research is to provide *intuition* to an algorithm, apart from what they are trained to *know* in a supervised manner. We present a novel intuition learning framework that learns to perform a task completely from unlabeled data. The proposed framework uses a continuous state reinforcement learning mechanism to learn a feature representation and a data-label mapping function using unlabeled data. The mapping functions and feature representation are succinct and can be used to supplement any supervised or semi-supervised algorithm. The experiments on the CIFAR-10 database show challenging cases where intuition learning improves the performance of a given classifier.

Keywords Reinforcement learning · Unsupervised classification · Intuition

1 Introduction

Intuition refers to knowledge acquired without inference and/or the use of reason [25]. Philosophically, there are several definitions for intuition and the most popularly used one is "Thoughts that are reached with little apparent effort, and typically without conscious awareness" [11] and is considered as the opposite of a rational process.

A. Sankaran
IBM Research, Bangalore, India
e-mail: anussank@in.ibm.com

M. Vatsa · R. Singh (✉)
Department of Computer Science and Engineering, IIIT Delhi, New Delhi, India
e-mail: rsingh@iiitd.ac.in

M. Vatsa
e-mail: mayank@iiitd.ac.in

© Springer Nature Switzerland AG 2020
R. Singh et al. (eds.), *Domain Adaptation for Visual Understanding*,
https://doi.org/10.1007/978-3-030-30671-7_8

From a machine learning perspective, training a supervised classifier is a rational process where it is trained with labeled data allowing it to learn a decision boundary. Also, traditional unsupervised learning methods do not map the learnt patterns to their corresponding class labels. Semi-supervised approaches bridge this gap by leveraging unlabeled data to better perform supervised learning tasks. However, the final task (say, classification) is performed only by a supervised classifier using labeled data with some additional knowledge from unsupervised learning. The notion of intuition would mean that the system performs tasks using only unlabeled data without any supervised (rational) learning. In other words, intuition is a context-dependent *guesswork* that can be incorrect at certain times. In a typical *learning* pipeline, the concept of intuition can be used for a variety of purposes starting from training data selection up to and including decision-making. Heuristics are the simplest form of intuition that bypass or is used in conjunction with rational decisions to obtain quick approximate results. For example, heuristics can be used in (1) choosing the new data points in an online active learning scenario [6], (2) for feature representation [7], (3) feature selection [10], or (4) choice of classifier and its parameters [4].

Table 1 shows the comparison of existing popular machine learning paradigms. Supervised learning attempts to learn an input–output mapping function on a feature

Table 1 Comparison of existing popular machine learning paradigms along with the proposed intuition learning paradigm

Paradigm	Input data	Learnt function	Comments
Supervised [3]	<data, label>	data-label mapping	
Unsupervised [3]	<data>	data clusters	
Semi-supervised [5]	<data, label>, unlabeled data	data-label mapping	unlabeled data follow the same distribution
Reinforcement [14]	reward function (or value)	state, action policy	need a teacher to provide reward
Active [24]	<data, label>	data-label mapping, new data selection	need human annotator (Oracle) or expert algorithm to provide labels for new data
Transfer [22]	<sourceData, sourceLabel>, <targetData, targetLabel>	targetData—targetLabel mapping	transfer can be data instances, classification parameters, or features
Imitation [18]	sourceData, sourceData-sourceLabel mapping	targetData—targetLabel mapping	need a teacher to provide reward
Self taught [23]	<data, label>, unlabeled data	data-label mapping	unlabeled data need not follow the same distribution and label as data
Deep learning [2]	<data, label>, unlabeled data	data-label mapping	complex architecture to learn robust data representations
Intuition	data, unlabeled data, reward function (or value)	data-label mapping	unlabeled data need not follow the same distribution, need a reward function

space using a set of labeled training data. Transfer learning aims to improve the target learning function using the knowledge in source (related) domain and source learning tasks [22]. Many types of knowledge transfer such as classification parameters [17], feature representations [9], and training instances [12] have been tested to improve the performance of supervised learning tasks. Semi-supervised learning utilizes additional knowledge from unlabeled data, drawn from the same distribution and having the same task labels as the labeled data. Many of the research works have focused on unsupervised feature learning, i.e., to create a feature subspace using the unlabeled data, to which the labeled data can be projected to obtain a new feature representation [5]. In 2007, Raina et al. [23] proposed a framework termed as "Self-taught learning" to create the generic feature subspace using sparse autoencoders irrespective of the task labels. Self-taught learning dismisses the same class label assumption of semi-supervised learning and forms a generic high-level feature subspace from the unlabeled data, where the labeled data can be projected.

As shown in Fig. 1, we postulate a framework of supplementing *intuition decisions* at the decision level to a supervised or semi-supervised classifier. The decisions drawn by the reinforcement learning block in Fig. 1 are called *intuition* because they are learnt only using the unlabeled data with an indirect reward from a teacher. Existing algorithms, broadly, require training labels for building a classifier or borrows the classifier parameters from an already trained classifier. Direct or indirect training is not always possible as obtaining data labels are very costly. To address this challenge, we propose a novel paradigm for unsupervised task performance mechanism learnt from cumulative experience. Intuition is modeled as a learning framework, which provides the ability to learn a task completely from unlabeled data. By using continuous state reinforcement learning as a classifier, the framework learns to perform the classification task without the need for explicit labeled data. Reinforcement learning helps in adapting a randomly initialized feature space to the specific task at hand, where a parallel supervised classifier is used as a teacher. As the proposed framework is able to learn a mapping function from the input data to the output class labels, without the requirement for explicit training, it functions similar to human intuition and we term this approach as *Intuition Learning*.

1.1 Research Contributions

This research proposes a novel intuition learning framework to enable algorithms learn a specific classification or regression task completely from unlabeled data. The major contributions of this research are as follows:

- A continuous state reinforcement learning-based classification framework is proposed to map input data to output class label, without the explicit use of training.
- A residual Q-learning-based function approximation method for learning the feature representation of task-specific data. A novel reward function which does not

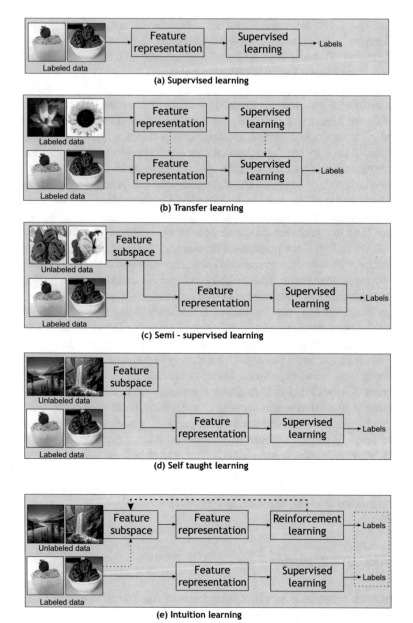

Fig. 1 Comparing the different learning paradigms such as supervised, semi-supervised, and transfer learning with the proposed intuition learning paradigm. Intuition learning transfer the knowledge to perform classification from unlabeled data using reinforcement learning

require class labels is designed to provide feedback to the reinforcement-based classification system.
- A context-dependent addition framework is proposed, where the result of the intuition framework can be supplemented based on the confidence of the trained supervised or semi-supervised mapping function.

2 An Intuition Learning Algorithm

The basic idea of the proposed intuition learning framework is presented in Fig. 2. Given a large set of unlabelled data, different kinds of feature representations are extracted to describe the data, irrespective of the task in hand. To further leverage the knowledge interpretation from unlabeled data, a continuous state reinforcement learning mechanism is used to perform the given classification task. As reinforcement is a continuous learning process, using a reward-based feedback mechanism, the classification task improves with time. The reinforcement learning, on one hand acts as a classifier, while on the other hand continuously adapts the feature representation with respect to the given task. Thus, given multiple tasks, the proposed intuition learning framework can adapt the generic feature space to be consistent with the corresponding task.

Let $\{(I_l^{(1)}, y^{(1)}), (I_l^{(2)}, y^{(2)}), \ldots, (I_l^{(m)}, y^{(m)})\}$ be the set of m labeled training data drawn i.i.d. from a distribution D. The labeled data are represented as $\{(x_l^{(1)}, y^{(1)}), (x_l^{(2)}, y^{(2)}), \ldots, (x_l^{(m)}, y^{(m)})\}$, where $x_l^{(i)} \in R^n$ is the feature representation of the

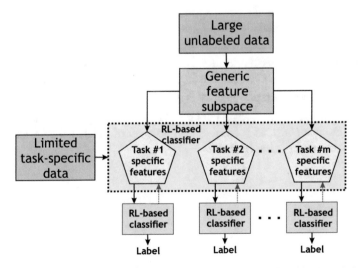

Fig. 2 A block diagram outlining on how a feature space can be adapted using reinforcement learning algorithm with feedback from a supervised classifier trained on limited task-specific data

data $I_l^{(i)}$ and $y^{(i)} \in [1, 2, \ldots, C]$ denotes the class label corresponding to $x_l^{(i)}$. Let the set of unlabeled data be $\{I_u^{(1)}, I_u^{(2)}, \ldots, I_u^{(p)}\}$, where the subscript u represents that they are unlabeled data. Contrary to self-taught learning [23], we do not assume that the labeled and unlabeled data should be drawn from the same distribution D or have the same class labels, however, they should be derived from the same modality. Given a set of labeled and large unlabeled data, the aim of intuition learning is to learn a hypothesis $h' : (X \rightarrow R) \in [1, 2, \ldots, C]$ that predicts the labels for a given input representation of data drawn. However, the hypothesis h' is learnt without the direct use of labels $y^{(i)}$ and is used as a supplement for the hypothesis h learnt using $(x_l^{(1)}, y^{(1)})$ in a supervised (or semi-supervised) manner.

2.1 Adapting Feature Representation

From a large set of unlabeled data, many different kind of feature representations are extracted. Each representation may correspond to a different property of the data that we try to capture. For image data, the features could be color, texture, and shape while for text data, the features could be n-grams, bag-of-words, and word embeddings. The features can also be a set of different color features or set of hierarchical n-grams. If the large set of unlabeled data is seen as the world (or the universal set), the features are the different observations made by the algorithm from the world. Similar to human intuition, the set of feature representations extracted are task-independent, and later depending on the learning task a subset of these features could be dominantly used. This task-independent feature space is similar to the human intuition learnt by observing the environment.

Figure 3 provides a detailed description of the proposed intuition learning framework. From the set of unlabeled data I_u, we extract r different kinds of feature representations, $\{X_{u_1}, X_{u_2}, \ldots, X_{u_r}\}$, where $X_{u_i} = \{x_{u_i}^{(1)}, x_{u_i}^{(2)}, \ldots, x_{u_i}^{(p)}\}$, where $x_{u_i}^{(j)} \in R^{n_i}$. For every feature representation $q \in [1, 2, \ldots, r]$, we cluster the representation $[x_{u_q}^{(1)}, x_{u_q}^{(2)}, \ldots, x_{u_q}^{(p)}]$ into C clusters[1] using k-means clustering. The centroid of each cluster for the ith feature representation is given as $[z_{u_{(i)}}^1, z_{u_{(i)}}^2, \ldots, z_{u_{(i)}}^C]$. This feature collection of $[z_{u_{(q)}}^1, z_{u_{(q)}}^2, \ldots, z_{u_{(q)}}^C]$, for $q = [1, 2, \ldots, r]$ is called as *Intuition-based Feature Subspace (IFS)*, as it clusters the entire set of unlabeled data into groups, based on every observation (feature).

---

[1]The best adaption results are obtained when we fix C to be the number of classes we have in the learning task.

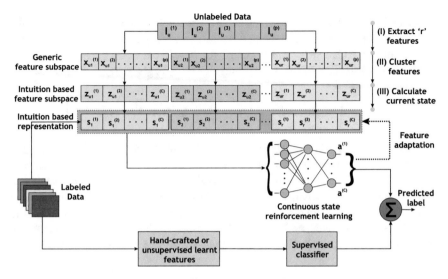

Fig. 3 Overall scheme of the proposed intuition learning algorithm that aids a supervised classifier

2.2 Classification Using Reinforcement Learning

For a given set of m labeled training data, $\{I_l^{(1)}, I_l^{(2)}, \ldots, I_l^{(m)}\}$, the set of r features (as used for the unlabeled data) are extracted as $[x_{l_q}^{(1)}, x_{l_q}^{(2)}, \ldots, x_{l_q}^{(m)}]$, where $q = [1, 2, \ldots, r]$. The extracted features are then projected onto the Intuition-based Feature Subspace (*IFS*) by calculating the distance of features from the corresponding cluster centroids shown as,

$$s_q^{(i)} = ||x_{l_q}^{(i)} - z_{u_q}^{(j)}||_2 \tag{1}$$

for $j = [1, 2, \ldots, C]$, $q = [1, 2, \ldots, r]$, and $i = [1, 2, \ldots, m]$. The representation of the data i is given by concatenating the distances corresponding to all the features,

$$s^{(i)} = [s_1^{(i)}, s_2^{(i)}, \ldots, s_r^{(i)}] \tag{2}$$

The obtained representation is succinct with a fixed length dimension of $rC \times 1$, where r is the number of different feature types extracted and C is the number of clusters. In essence, every value represents the distance from a cluster centroid. Also, in a typical semi-supervised (or self-taught) learning scheme, the mapping between intuition based representation and the output class labels, $\{(s^{(1)}, y^{(1)}), (s^{(2)}, y^{(2)}), \ldots, (s^{(m)}, y^{(m)})\}$ is learnt in a supervised manner. However, in the proposed intuition learning, we attempt to learn the data-label mapping without using the class labels, using reinforcement learning. The aim of reinforcement learning is to learn an action policy $\pi : s \to a$, where $s \in S$ is the current state of the system and a is the action performed in that state. As the setup involves a continuous state environment, the optimal action policy is learnt using a model free, off-policy Temporal Difference (TD) algorithm called *Q-learning*, where $Q(s, a)$-value denotes the effectiveness of a state-action pair. The $TD(0)$ Q-learning algorithm is given by,

$$Q(s_t, a) = Q(s_t, a) + \alpha \left[r_t + \gamma \max_{a'} Q(s_{t+1}, a') - Q(s_t, a) \right] \quad (3)$$

where $r_t \in R^n$ is the immediate reward obtained for performing action a in state s_t, $\gamma \in [0, 1]$ is the factor with which the future rewards are discounted and $\alpha \in [0, 1]$ is the learning rate. In our problem, reinforcement learning is formulated as a classification problem, where *IFS* is the current state s and action a is the output label to be predicted, the policy π learns the data-label relation for the given data. Due to the large, probabilistic, and continuous definition of the space s, the Q-values are approximated using a universal function approximation, i.e., a neural network [26].

$$Q(s, a) = \psi(s, a, \theta) = \sum_i \phi_i(s, a).\theta_i = \phi^T(s, a).\theta \quad (4)$$

where ϕ is the approximation function. Using residual Q-learning algorithm [21], the free parameters θ are updated as follows:

$$\theta_{t+1} = \theta_t + \alpha.\psi.\Delta\psi \quad (5)$$

$$\theta_{t+1} = \theta_t + \alpha \left[r_t + \gamma \max_{a'} Q(s_{t+1}, a') - Q(s_t, a) \right] \\ \times \left[\beta\gamma \frac{\partial}{\partial\theta} \max_{a'} Q(s_{t+1}, a') - \frac{\partial}{\partial\theta} Q(s_t, a) \right] \quad (6)$$

where β is a weighting factor called the Bellman residual. Baird [1] guaranteed the convergence of the above approximate Q-learning function, the details of which are skipped for the sake of brevity. $\epsilon-$ exploration strategy is adopted, where, in every state a random action is preferred with a probability of ϵ. As observed in [16], *"the crucial factor for a successful approximate algorithm is the choice of the parametric approximation architecture and the choice of the projection (parameter adjustment) method(s)"*. The choice of reward function employed is highly important and directly implies the effectiveness of adaption, which is explained in the next section.

2.3 Design of Reward Function

The Intuition-based Feature Subspace (IFS) is defined by the cluster centroid points obtained using unlabeled data for every feature q as $[z_{u_q}^{(1)}, z_{u_q}^{(2)}, \ldots, z_{u_q}^{(C)}]$, where

$q = [1, 2, \ldots, r]$. This space provides an organized definition of how the entire set of unlabeled data is observed and inferred. From the various features of the labeled training data $[(x_{l_q}^{(1)}, y^{(1)}), (x_{l_q}^{(2)}, y^{(2)}), \ldots, (x_{l_q}^{(m)}, y^{(m)})]$, where $q \in [1, 2, \ldots, r]$, the centroid points for every feature and every class are calculated as, $[z_{l_q}^{(1)}, z_{l_q}^{(2)}, \ldots, z_{l_q}^{(C)}]$, where $q = [1, 2, \ldots, r]$. This space, called the *Labeled data Feature Subspace (LFS)*, formed by these centroid points provide us the inference of the particular learning task to be performed. It is to be noted that:

- Apart from unlabeled data, every labeled training data (and even testing data) gets incrementally added to the *IFS*, as the observed data affects the overall understanding of features.
- The aim of incremental learning is to shape the IFS as close as possible to LF while learning the feature-label mapping using reinforcement learning.

The incremental update of the IFS happens for the ith training example belonging to jth class, as shown in the following equation:

$$z_{u_q}^{(j)} = z_{u_q}^{(j)} + \left(\frac{x_{l_q}^{(i)} - z_{u_q}^{(j)}}{n_q^j} \right) \tag{7}$$

for $q = [1, 2, \ldots, r]$, where n_q^j is the number of data points in the jth cluster for qth feature. Further, to make effective learning from this incremental update, the reward function is defined as a function of the distance between the current IFS and LFS, as follows:

$$r_t = \left(||z_{u_q,t}^{(j)} - z_{l_q}^{(j)}||_2 \right)^{-1} \tag{8}$$

for $q = [1, 2, \ldots, r]$, $j = [1, 2, \ldots, C]$ at a given time t.

2.4 Context-Dependent Addition Mechanism

Intuition learning framework acts as a supplement to (and not complementing) supervised learning. The need for intuition arises only when the confidence of supervised learner falls below a particular threshold. Therefore, a context-dependent mechanism is designed to leverage supervised learning using intuition only when required. For given labeled training data $\{I_l^{(1)}, I_l^{(2)}, \ldots, I_l^{(m)}\}$, some handwritten or unsupervised features are extracted, $\{(x_l^{(1)}, y^{(1)}), (x_l^{(2)}, y^{(2)}), \ldots, (x_l^{(m)}, y^{(m)})\}$ and a supervised model is learnt, $H_s : \left(x_l^{(i)} \to \hat{y}_s \right)$. Based on the supervised learning algorithm, the classification confidence is computed for the ith data point and is given as $conf_s^{(i)} = [cs_1^{(i)}, cs_2^{(i)}, \ldots, cs_C^{(i)}]$. The mechanism to calculate the classification confidence depends on the supervised learning model used. Similarly, the intuition learning can be represented as $H_{int} : \left(s^{(i)} \to \hat{y}_{int} \right)$ and the classification confidence is the output of the last layer of the value function approximation neural architecture, given as $conf_{int}^{(i)} = [cint_1^{(i)}, cint_2^{(i)}, \ldots, cint_C^{(i)}]$. A label switching mechanism is performed to give the final predicted label, $\hat{y}$, as follows:

$$\hat{y} = \begin{cases} \hat{y}_s, & \Delta > th \\ \hat{y}_{new}, & \text{otherwise} \end{cases} \tag{9}$$

where th is the threshold for using intuition and the condition for context Δ is calculated as follows:

$$\Delta = \max_j \left(cs_j^{(i)}\right) - \max_{l \neq j} \left(cs_l^{(i)}\right) \tag{10}$$

In such cases where intuition is used to boost the confidence of supervised classifier the new label is computed as follows:

$$cnew_k^{(i)} = \lambda.cs_k^{(i)} + (1 - \lambda).cint_k^{(i)} \tag{11}$$

$$\hat{y}_{new} = \arg \max_j \left(cnew_j^{(i)}\right) \tag{12}$$

where λ is the trade-off parameter between intuition and supervised learning. Thus, in simple words, we add the feeling of intuition to an algorithm. The entire approach is summarized as an algorithm in Algorithm 1.

Algorithm 1 Intuition Learning Algorithm

1: **Input:** Labeled Data: $\{(I_l^{(1)}, y^{(1)}), (I_l^{(2)}, y^{(2)}), \ldots (I_l^{(m)}, y^{(m)})\}$, Unlabeled Data: $\{I_u^{(1)}, I_u^{(2)}, \ldots I_u^{(p)}\}$, $maxNumberOfEpochs$

2: **repeat**
3:    **for** $i = 1$ **to** m **do**   ▷ Extract r different types of features
4:        $\{[x_{u_1}^{(1)}, x_{u_1}^{(2)}, \ldots, x_{u_1}^{(p)}], [x_{u_2}^{(1)}, x_{u_2}^{(2)}, \ldots, x_{u_2}^{(p)}], \ldots,$ $[x_{u_r}^{(1)}, x_{u_r}^{(2)}, \ldots, x_{u_r}^{(p)}]\} \leftarrow$
         $\{I_u^{(1)}, I_u^{(2)}, \ldots I_u^{(p)}\}$ ▷ Cluster data into C groups based on each feature
5:        $[z_{u_q}^1, z_{u_q}^2, \ldots, z_{u_q}^C], \forall q = [1, 2, \ldots, r]$ ▷ Compute the current state
6:        $s_q^{(i)} = ||x_{l_q}^{(i)} - z_{u_q}^j||_2, \quad \forall j = [1, 2, \ldots, C], \quad \forall q = [1, 2, \ldots, r], \quad \forall i = [1, 2, \ldots, m]$
         ▷ Approximate the current state Q-value
7:        $Q(s, a) = \psi(s, a, \theta) = \sum_i \phi_i(s, a).\theta_i$ ▷ Update θ
8:        $\theta_{t+1} = \theta_t + \alpha.\psi.\Delta\psi$ ▷ Compute reward value
9:        $r_t = \left(||z_{u_q,t}^j - z_{l_q}^j||_2\right)^{-1} \forall q = [1, 2, \ldots, r], \forall j = [1, 2, \ldots, C]$
10:    **end for**
11: **until** $maxNumberOfEpochs$ or $\Delta\psi < thresh$ ▷ Test phase
12: For $I_t^{(i)}$, calculate $\beta, \hat{y}_s, \hat{y}_{new}$
13: **if** $\Delta > thresh$ **then**
14:    $\hat{y} \leftarrow \hat{y}_s$
15: **else**
16:    $\hat{y} \leftarrow \hat{y}_{new}$, as shown in Eqs. 11, 12.
17: **end if**

---

3 Experimental Analysis

3.1 *Dataset*

The proposed intuition learning algorithm is applied for 10-class classification problem using the CIFAR-10 database [15]. The database contains 60,000 color images labeled, each of size 32×32 pertaining to 10 classes (i.e., 6,000 images per class). There are 50,000 training images and 10,000 test images. The data set contains small size images, leading to limited and noisy information content and it provides the most relevant case study to demonstrate the effectiveness of the proposed paradigm. The STL-10 database [7] is used as the unlabeled image data set having one million colored images of size 96×96. As shown in Table 2 six different feature representations are extracted from the images. These features comprehensively comprise the various types of features that could be extracted from image data. For all the experiments, five times random cross-validation is performed and the best model accuracy is reported for all the experiments. Sample images from CIFAR-10 and STL-10 datasets are shown in Fig. 4.

3.2 *Interpreting Intuition-Based Feature Subspace*

The primary aim of the approach is to construct the feature subspace completely from unlabeled data and to adapt it to a specific learning task. Figure 5 shows the clusters of entire unlabeled data corresponding to every feature extracted. The concatenation of the feature spaces put together in Fig. 5a represents the *IFS*. Figure 5b shows the adapted task-specific feature subspace after performing 300 epochs of learning with the given labeled data. Figure 5c shows the amount of update in the cluster after adding an image, by calculating the dissimilarity between the cluster centroid, before and after the addition of the image. Cluster dissimilarity is calculated for the $r - th$ feature representation as follows:

Table 2 Details of different features extracted from the image data

Type	Feature	Dimension	Parameters
Color	Color harris [27]	10×2	$\sigma_g = 1.5, \sigma_a = 5$
Color	Color autocorrelograms [13]	64×1	Quantization level, $m = 64$
Local texture	Local binary pattern (LBP) [19]	59×4	$N = 8, R = 1$
Global texture	GIST [20]	512×1	$n_\theta = 8, n_{block} = 4$
Saliency	Region covariances [8]	32×32	$r = 3, \sigma = 1.2, m = 1/10$
Shape	Multilayer autoencoder [28]	10×1	size = $[10, 110, 110, 340, 340, 1024, 1024]$

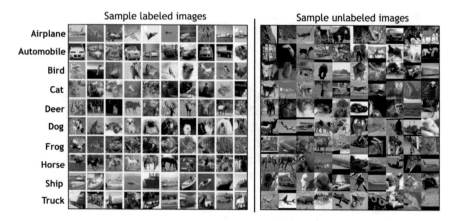

Fig. 4 Sample set labeled images from CIFAR-10 database and unlabeled images from STL-10 database

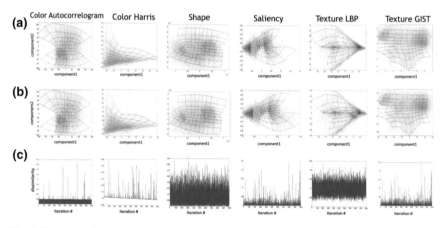

Fig. 5 Image showing the data clusters for each of the extracted feature and grid depicts the cluster density at local regions. **a** shows the *IFS* of all the unlabeled data, **b** shows the adapted task-specific feature space after 300 epochs of learning, and **c** shows the amount of change happening in the cluster after the addition of an image. Best viewed in color

$$C_{dis} = \sum_{j=1}^{C} \frac{1}{D_j} \cdot ||z_{u_r}^{(j)}|_{(t+1)} - z_{u_r}^{(j)}|_{(t)}||_2 \tag{13}$$

where D_j is the density of the jth cluster. It can be visually observed from the plot that, shape, gist, and LBP feature spaces are updated (learns) after the addition of each image, indicating that these features contribute more towards the classification task. However, both color Harris and autocorrelogram features are not much updated by the training data.

3.3   Performance Analysis

It is to be noted that intuition learning framework is used to supplement any supervised or semi-supervised learning mechanism. In this research, we show the results in the following scenario:

1. Using two supervised learning algorithms (backpropagation neural network and multi-class SVM) with Uniform Circular Local Binary Pattern (UCLBP) [19] as features. Labeled data, $[(x_{l_q}^{(1)}, y^{(1)}), (x_{l_q}^{(2)}, y^{(2)}), \ldots, (x_{l_q}^{(m)}, y^{(m)})]$, from CIFAR-10 is used to train the supervised algorithms.
2. Using a semi-supervised learning algorithm, with neural network as classifier and UCLBP features trained on CIFAR-10 dataset. The semi-supervised algorithm used for comparison is one approach for self-taught learning [23], with unlabeled data from STL-10 dataset, $\{(s^{(1)}, y^{(1)}), (s^{(2)}, y^{(2)}), \ldots, (s^{(m)}, y^{(m)})\}$.
3. Using a intuition learning framework only, having the intuition-based task-specific feature representation combined with a continuous state reinforcement learning (Q-learning) in Eq. 4 for classification.
4. Using a supervised intuition framework, where the output of the supervised learning algorithm and the intuition learning framework is combined using the context-dependent addition mechanism in Eq. 12.
5. Using a semi-supervised intuition framework, where the output of the semi-supervised learning algorithm and the intuition learning framework is combined using the context-dependent addition mechanism.

The optimized values of various parameters used in our framework are as follows: $\alpha = 0.99$, $\gamma = 0.95$, $\beta = 0.2$, $th = 0.9$, $\lambda = 0.5$, and $\epsilon = 0.05$. Preprocessing of features is done using z-score normalization. All the experiments are performed on a Intel Xeon $E5 - 2640\ 0$, $2.50\,\text{GHz}$, $64\,\text{GB RAM}$ server.

As already discussed, intuition has a better significance in challenging problems with limited training data. Tables 3, 4, and 5 show the performance of the proposed intuition learning in comparison with other learning methods, by varying the train-

Table 3 The performance accuracy (%) of supervised intuition learning is compared with supervised (neural network) and semi-supervised (self-taught) learning methods. The significance of intuition is studied by varying the amount of available training data. 5 times random cross-validation is performed and the best models performance is reported

Training size	Supervised	Semi-supervised	Supervised intuition
10000	**38.90**	29.64	36.19
5000	**37.13**	24.34	34.78
3000	14.12	17.07	**25.65**
1000	10.34	16.54	**19.61**

Table 4 The influence of supplementing intuition to supervised and semi-supervised algorithm is shown by improvement in the performance accuracy (%)

Training size	Intuition	Intuition supervised	Intuition semi-supervised
10000	12.11	36.19	29.21
5000	10.00	34.78	23.85
3000	10.00	**25.65**	**22.49**
1000	08.99	**19.61**	**20.53**

Table 5 The performance accuracy (%) of supervised and supervised intuition framework using SVM classifier is studied

Training size	Supervised	Intuition supervised
10000	**44.57**	41.83
5000	**43.21**	40.77
3000	13.56	**17.48**
1000	06.19	**09.78**

(a) AL: Ship SL: Truck IL: Ship   AL: Cat SL: Horse IL: Cat   **(b)** AL: Horse SL: Horse IL: Cat   AL: Truck SL: Truck IL: Frog

Fig. 6 Examples of **a** success and **b** failure cases of the proposed intuition learning. AL = actual ground truth label, SL = label predicted by the supervised neural network learner, and IL = label predicted when intuition is combined with supervised neural network learner

ing size as parameter.[2] It can be observed that with enough training data, supervised algorithms (both neural network and SVM) yield the best classification performance. However, with decrease in the size of training data, the performance of all the three algorithms, supervised, semi-supervised, and intuition learning reduces. The results show that in such a scenario, incorporating intuition with supervised or semi-supervised algorithm yields improved results. This supports our hypothesis that adding intuition would improve the performance from under challenging circumstances such as limited training data. Similarly from a human's perspective, under

---

[2]For a given training size, the same subset of images is used across all the classifiers to avoid any training bias.

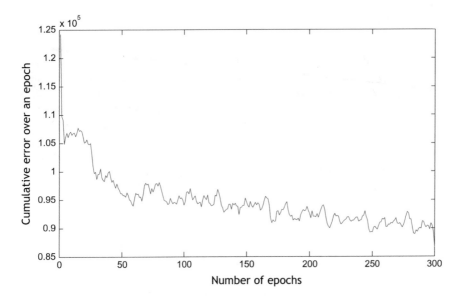

Fig. 7 A plot between the cumulative errors across each epoch empirically showing the learning effectiveness of the residual Q-learning performed in Eq. 6

the presence of all data and information, one may take correct decisions. However, when the background training data information is limited, intuition learning helps. Further, some key analysis are summarized below:

1. To study the effectiveness of residual learning in Eq. 6, training error over successive epochs is plotted, as shown in Fig. 7, for a training size of 10000. It can be observed that the training error gradually decreases and remains constant after 300 epochs, indicating that maximum training capacity has been achieved, with minimum training error.

2. The computation time required for intuition learning depends on the complexity of r features that are extracted. However, for one sample, under the assumption that the feature extraction happens off-line, the overall intuition decision and feature space can be generated in 0.082 s while the supervised decision can be taken in ~4 s on an average. This shows that intuition is much faster requiring little effort than supervised decision-making.

3. In Fig. 6, some success and failure example cases are shown where (a) intuition helps incorrectly classifying a data but supervised learning fails and (b) data was incorrectly classified because of intuition. As previously discussed, intuition can go wrong sometimes. Upon analyzing the first horse example in failure case (Fig. 6b), it is observed that horses are clustered more towards brown color in the autocorrelogram color feature space. However, as the horse shown in the images is white in color, it gets clustered along with cat and misclassified by intuition learning.

4 Conclusion

Inspired from human capabilities of instinct reasoning, this research presents a intuition learning framework that supplements a classifier for improved performance, especially with limited training data. Intuition is modeled as a continuous state reinforcement learning, that adapts to a particular task using large amount of unlabeled data and limited task-specific data. The performance of intuition is shown in a 10 class image classification problem, in comparison with supervised, semi-supervised, and reinforcement learning. The results indicate that the application of intuition improves the performance of the classifier with limited training data.

References

1. Baird L (1995) Residual algorithms: reinforcement learning with function approximation. In: ICML, pp 30–37
2. Bengio Y (2009) Learning deep architectures for AI. Found Trends Mach Learn 2(1):1–127
3. Bishop CM, Nasrabadi NM (2006) PRML, vol 1
4. Cavalin PR, Sabourin R, Suen CY (2012) LoGID: an adaptive framework combining local and global incremental learning for dynamic selection of ensembles of HMMs. PR 45(9):3544–3556
5. Chapelle O, Schölkopf B, Zien A et al (2006) Semi-supervised learning, vol 2. MIT Press, Cambridge
6. Chu W, Zinkevich M, Li L, Thomas A, Tseng B (2011) Unbiased online active learning in data streams. In: ACM SIGKDD, pp 195–203
7. Coates A, Ng AY, Lee H (2011) An analysis of single-layer networks in unsupervised feature learning. In: ICAIS, pp 215–223
8. Erdem E, Erdem A (2013) Visual saliency estimation by nonlinearly integrating features using region covariances. J Vis 13(4)
9. Evgeniou A, Pontil M (2007) Multi-task feature learning. In: NIPS, vol 19, p 41
10. Guyon I, Elisseeff A (2003) An introduction to variable and feature selection. JMLR 3:1157–1182
11. Hogarth RM (2001) Educating intuition. University of Chicago Press, Chicago
12. Huang J, Gretton A, Borgwardt KM, Schölkopf B, Smola AJ (2006) Correcting sample selection bias by unlabeled data. In: NIPS, pp 601–608
13. Huang J, Kumar SR, Mitra M, Zhu WJ, Zabih R (1997) Image indexing using color correlograms. In: CVPR, pp 762–768
14. Kaelbling LP, Littman ML, Moore AW (1996) Reinforcement learning: a survey. arXiv:cs/9605103
15. Krizhevsky A, Hinton G (2009) Learning multiple layers of features from tiny images. Master's thesis, Department of Computer Science, University of Toronto
16. Lagoudakis MG, Parr R (2003) Least-squares policy iteration. JMLR 4:1107–1149
17. Lawrence ND, Platt JC (2004) Learning to learn with the informative vector machine. In: ICML, pp 65–72
18. Natarajan S, Joshi S, Tadepalli P, Kersting K, Shavlik J (2011) Imitation learning in relational domains: a functional-gradient boosting approach. In: IJCAI, pp 1414–1420
19. Ojala T, Pietikainen M, Maenpaa T (2002) Multiresolution gray-scale and rotation invariant texture classification with local binary patterns. IEEE TPAMI 24(7):971–987
20. Oliva A, Torralba A (2001) Modeling the shape of the scene: a holistic representation of the spatial envelope. IJCV 42(3):145–175

21. Paletta L, Pinz A (2000) Active object recognition by view integration and reinforcement learning. Robot Auton Syst 31(1):71–86
22. Pan SJ, Yang Q (2010) A survey on transfer learning. IEEE TKDE 22(10):1345–1359
23. Raina R, Battle A, Lee H, Packer B, Ng AY (2007) Self-taught learning: transfer learning from unlabeled data. In: ICML, pp 759–766
24. Settles B (2010) Active learning literature survey. Technical report, Computer Sciences Technical Report 1648, University of Wisconson, Madison
25. Simpson JA, Weiner ES et al (1989) The Oxford english dictionary, vol 2. Clarendon Press, Oxford
26. Sutton RS, Barto AG (1998) Reinforcement learning: an introduction. Cambridge University Press, Cambridge
27. Van De Weijer J, Gevers T, Smeulders AW (2006) Robust photometric invariant features from the color tensor. IEEE TIP 15(1):118–127
28. Vincent P, Larochelle H, Bengio Y, Manzagol PA (2008) Extracting and composing robust features with denoising autoencoders. In: ICML, pp 1096–1103

Alleviating Tracking Model Degradation Using Interpolation-Based Progressive Updating

Xiyu Kong, Qiping Zhou, Yunyu Lai, Muming Zhao and Chongyang Zhang

Abstract Recently, Correlation Filter (CF)-based methods have demonstrated excellent performance for visual object tracking. However, CF-based models often face one model degradation problem: With low learning rate, the tracking model cannot be updated as fast as the large-scale variation or deformation of fast motion targets; As for high learning rate, the tracking model is not robust enough against disturbance, such as occlusion. To enable the tracking model adapt with such variation effectively, a progressive updating mechanism is necessary. In order to exploit spatial and temporal information in original data for tracking model adaptation, we employ an implicit interpolation model. With motion-estimated interpolation using adjacent tracking frames, the obtained intermediate response map can fit the learning rate well, which will effectively alleviate the learning-related model degradation. The evaluations on the benchmark datasets KITTI and VOT2017 demonstrate that the proposed tracker outperforms the existing CF-based models, with advantages regarding the tracking accuracy.

Keywords Visual tracking · Correlation filter · Progressive updating

1 Introduction

Visual tracking is one of the most challenging tasks in computer vision. It has attracted a lot of interests from numerous researchers for its wide applications in diverse areas such as video surveillance, vehicle autonomy, and video analysis.

Recently, Correlation Filter (CF) based methods [4, 11, 16, 17], have become a popular approach to deal with tracking task due to the advantage of computational efficiency. CF-based tracking process can be decomposed into three steps: training

X. Kong · M. Zhao · C. Zhang (✉)
Institute of Image Communication and Network Engineering,
Shanghai Jiao Tong University, Shanghai 200240, China
e-mail: sunny_zhang@sjtu.edu.cn

Q. Zhou · Y. Lai
State Grid Jiangxi Power Co. Ltd, Maintenance Branch, Nanchang 330096, China

© Springer Nature Switzerland AG 2020
R. Singh et al. (eds.), *Domain Adaptation for Visual Understanding*,
https://doi.org/10.1007/978-3-030-30671-7_9

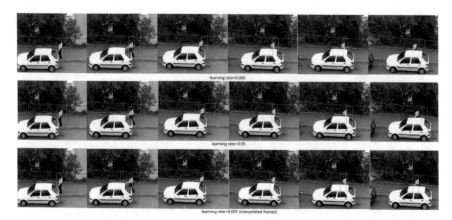

Fig. 1 Illustration of tracking failure under different learning rates: The first row illustrates tracking failure results from relatively low learning rate (0.005), the tracking model is not updated fast enough in order to match with fast-changing target; The second row illustrates tracking failure results from relatively high learning rate (0.05), which makes tracking model update so fast that the disturbance, such as occlusion, is taken as the main part of the target, and thus wrong result is got; The third row shows accurate tracking result based on the proposed method using progressive updating, which has the same learning rate as the first row

the correlation filter by performing circular sliding window operation to periodically extended training samples [24]; detecting the target with the trained tracker in a new frame; updating the tracking model with the detection result. CF methods gain the advantage of computational efficiency based on a principle that training and detection samples must be extended through circular shifts. Although dense sampling provides sufficient feature from the data, cyclic shifts create a synthetic region of which the detection scores are only accurate near the center [13]. This leads to a restricted target search region at the detection step. For a target with fast motion, the CF-based trackers tend to generate tracking failure due to the restricted search region. Reference [8] From the perspective of model updating, tracking failure of these approaches can be divided into two situations: either with a low learning rate the tracking model updates too slow to catch up with variation of fast moved targets, or with a higher learning rate the tracking model is not robust enough against disturbance, such as occlusion (Fig. 1).

Tracking objects with occlusion (see Fig. 2) or fast motion (see Fig. 3) demands a robust tracking model capable of adapting with large search region and visual features variance. To alleviate effects brought by fast-moving objects, we propose a progressive updating mechanism based on motion-estimated interpolation method (MEINT) to update tracking model with transitional information. We argue that this non-model specific technique can provide intermediate spatial-temporal information for model training and detection process without relying on delicate learning rate fine-tune.

Based on this intuition, we propose a novel tracking model to alleviate model degradation situation especially with fast-moving targets. One motion-estimated

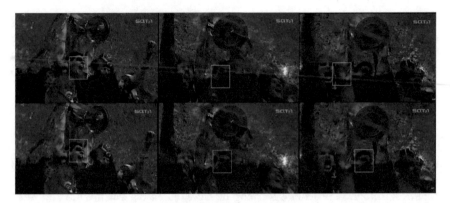

Fig. 2 Occlusion scenario in soccer1 of vot2017. The upper sequence is the result from DCF-CSR tracker. The lower sequence is the result from proposed method

Fig. 3 Fast motion/illumination change scenario in drone1 of vot2017. The upper sequence is DCF-CSR tracker's result. The lower sequence is the result from proposed method

interpolation (MEINT) block is developed in this work, which is used to generate interpolated response with adjacent tracking frames, and thus the model can be updated smoothly using the intermediate information. By adapting original data with spatial and temporal information, we propose a generic scheme which can be seamlessly integrated into those models to enhance the motion-adaptive capacity of the tracker itself.

Overall, our contribution to tracking model degradation objects are twofolds:

1. We propose a novel interpolation-based tracking model to pose the learning problem. With motion-estimated interpolation using adjacent tracking frames, the obtained intermediate response map can fit the learning rates well and thus learning-related model degradation can be reduced effectively.

2. One motion-estimated interpolation (MEINT)-based progressive updating mechanism is developed to obtain the accurate intermediate response. As a generic block, MEINT is complementary to existing approaches, and can be integrated with existing tracking models to enhance their motion-adaptive capacity.

2 Related Work

During the last decade, more and more valuable works have been developed and enabled great progress in the area of visual object tracking. ConvNets have enabled significant progress in feature learning recently, based on which more and more tracking methods are developed to replace handcrafted feature-based traditional models [1, 21]. However, most deep learning based methods [9, 22] suffer from high computational cost due to a large amount of workload during training process, which makes them not a suitable choice for real-time object tracking. Valmadre et al. [26] proposed an end-to-end representation for Correlation Filter-based tracking and achieved a speed of 45 fps, however, at a cost of accuracy which is even lower than some handcrafted feature based tracking methods [6] which have higher fps. Further more, the need of GPU also increases the cost of equipments, which limits the application of deep-based methods in real-time embedded systems.

In terms of non-deep learning methods, the discriminative correlation filters with fHOG [5] features have been popularized recently in the tracking community, starting with MOSSE tracker [4]. The kernelized correlation filter(KCF) [6] further investigated the computational efficiency results from circular structures.

However, the circular structure also leads to several problems known as boundary effect [12]. DCFs with limited boundaries (CFLB) [12], spatially regularized CF(SRDCF) [7], and background correlation filters (BACF) [10] all propose different approaches in order to get more important information from the training sample to achieve better tracking model. CFs with limited boundaries (CFLB) [12] was proposed to learn CFs with fewer boundary effects. However, it is merely based on pixel intensities, which was proved to be not good enough to express the patterns in the image in [11]. Spatially regularized CF (SRDCF) [7] was proposed to learn trackers with optimization objective altered in order to maintain more information around the center of the patches. Although this method has promising performance on prediction, it has a major disadvantage. Specifically, the regularized objective, without close-form solution, is costly to optimize. In addition, a set of parameters need to be carefully tuned to form the regularization weights. Failure to tune these hyperparameters may lead to poor tracking performance [10]. Background correlation filters (BACF) [10] were proposed to include more background information from the frame and generate negative training samples which will help the classifier distinguish background from the foreground. Despite its impressive performance, altering the structure of the training process leads to no closed-form solution and the authors resorted to ADMM methods to solve the problem, which leads to a speed of 35 fps.

3 Alleviating Model Degradation Using Interpolation-Based Progressive Updating

3.1 Revisiting of CF-Based Tracking Method

The CF-based tracking process can be mainly decomposed into three steps: training, detection, and model updating. A cosine window is applied on the features, in order to remove the boundary discontinuities. Due to the convenience brought by circular shifts, the convolution in the spatial domain can be expressed as an element-wise product in the frequency domain.

The parameters of the correlation filter contains: $\mathcal{F}(x)$ and $\mathcal{F}(\alpha)$, where $\mathcal{F}(\cdot)$ denotes the Fourier transformation. For the first frame, x is feature extracted from target region, while α is the filter parameter determined by label and auto-correlation of the target feature.

$$\mathcal{F}(\alpha) = \frac{\mathcal{F}(y)}{Corr(\mathcal{F}(x), \mathcal{F}(x))}. \tag{1}$$

where function $Corr(\cdot)$ denotes correlation in frequency domain. y is a Gaussian distributed label with the center at the target position.

The response in Fourier domain of the circular sample can be calculated directly with dot product. Let $Corr(\mathcal{F}(x), \mathcal{F}(z))$ denotes the correlation between the model feature $\mathcal{F}(x)$ and the feature extracted from detecting region, the response can be expressed as follows:

$$\mathcal{F}(R) = Corr(\mathcal{F}(x), \mathcal{F}(z)) \cdot \mathcal{F}(\alpha). \tag{2}$$

The new position is detected by searching for maximal response's location, which provides new training samples for new target model as stated in Eq. 1. The new training samples are subsequently used to update the model. Updating the model with new parameters also involves updating $\mathcal{F}(x)$ and $\mathcal{F}(\alpha)$. As mentioned above, the circular form of x leads to speed-up results from circular matrix: the correlation of circular matrix in Fourier domain can be expressed as dot product. The updating process can be expressed as follows:

$$\mathcal{F}(\alpha_t) = (1 - l) \cdot \mathcal{F}(\alpha_t) + l \cdot \mathcal{F}(\alpha_{t+1}). \tag{3}$$

$$\mathcal{F}(x_t) = (1 - l) \cdot \mathcal{F}(x_t) + l \cdot \mathcal{F}(x_{t+1}). \tag{4}$$

where l denotes the learning rate and t indexes the frame where the model is from. More detailed explanation of standard CF-based method can be found in [16].

The efficiency of CF is based on the circular shifts in the training samples and testing samples [7]. However, circular-shifted samples also cause problems: the cosine window applied to the data leads to restricted target search region and inaccurate data around the margin. More detailed explanation about problems result from circular shifts that can be found in [7].

For fast-moving objects, due to restricted detection area resulted from training process, the detection area from the previous frame fails to cover tracking object from the current frame, which leads to tracking failure. To alleviate effects brought by fast-moving objects, we propose a motion-estimated interpolation method (MEINT), which synthesis frames with transitional information.

3.2   Motion-Estimated Interpolation(MEINT) Based Progressive Updating

The equations in Sect. 3.1 still hold for the proposed method. Other than the original set of samples s, in order to alleviate model degradation, the proposed method progressively updates the tracking model for each interpolated frame, which provides a set of samples carrying spatial and temporal information.

Figure 4 illustrates the effect of progressive updating on frames with the depredating model, the model degradation is reduced due to the progressive updating based on interpolation of the frames.

Motion estimation(ME) is done by matching the same entities in two adjacent frames and calculating the motion vector field (MVF), relying on block matching [25] or on descriptor matching [23]. The core of various ME algorithms is block matching

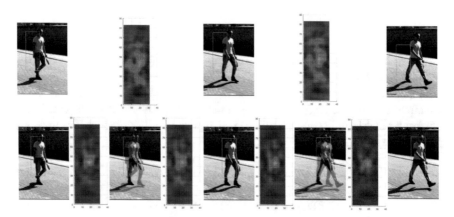

Fig. 4 Comparison between original frames and frames after progressive updating for response map and detection results. The response map is calculated from the cross-correlation between model and test sample. As shown in the first row, fast motion leads to model not robust enough, which leads to response map with multiple peak values, since it is possible for the highest peak value to be a disturbance, the tracking result might drift from the target. In the second row, interpolated frame provide transition information for training tracking model, which leads to distinct response map with single peak value and accurate tracking result

algorithm (BMA) [14] which has intuitive architecture and low complexity [19]. BMA is adopted here for lowering computation cost. The two adjacent frames are first divided into square blocks with length, and by matching blocks and recording the movement between two frames, the MVF is calculated.

For the block with center $\mathbf{c}$ and size $p \cdot p$, the block matching is done by minimizing the energy difference between blocks from two frames.

$$\underset{\delta_c}{\text{minimize}} \quad D(B(\mathbf{c}), B(\mathbf{c} + \delta_c))$$

$$\text{subject to} \quad \delta_c \in [-p, p]. \tag{5}$$

where $B(\mathbf{c})$ denotes the pixel block centered at $\mathbf{c}$, δ_c denotes estimated motion, and $D(\cdot)$ denotes the cost function chosen to describe the difference between blocks.

Its form depends on the terms of computational expense such as mean absolute difference (MAD) or mean square error(MSE). By minimizing the energy, the motion estimation of each block (δ_c) can be generated into MVF $\mathbf{u}(\mathbf{c})$, which is needed in the interpolation process. The MVF records the estimated movement for each block.

The interpolation method is adopted from [2]. In order to calculate the interpolated frame, the interpolated field shall be calculated first. From motion field $\mathbf{u}_t$ between frame f_t and frame f_{t+1}, let δ_t denote the time interval, the motion field on point $\mathbf{c}$ of the interpolated frame generated at time $t + \lambda \cdot \delta_t$ is expressed as:

$$\mathbf{u}_\lambda(\text{round}(\mathbf{c} + \lambda \mathbf{u}_t(\mathbf{c}))) = \mathbf{u}_t(\mathbf{c}). \tag{6}$$

For each hole $\mathbf{c}_{\text{hole}}$ in the interpolated motion field $\mathbf{u}_\lambda(\mathbf{c}_{\text{hole}})$, it is filled up with the nearest motion vector $\mathbf{u}_\lambda(\mathbf{c}_{\text{assigned}})$. The i-th interpolated frame between frame f_t and f_{t+1} is expressed as follows:

$$\begin{aligned} f_\lambda(\mathbf{c}) = (1 - \lambda) f_t(\mathbf{c} - \lambda \mathbf{u}_\lambda(\mathbf{c})) \\ + \lambda f_{t+1}(\mathbf{c} + (1 - \lambda)\mathbf{u}_\lambda(\mathbf{c})). \end{aligned} \tag{7}$$

3.3 Process of Progressive Updating Model

The process is illustrated in Algorithm 1, where emphasized part indicates the interpolation process based on motion estimation and black part indicates standard CF-based tracking process, $\mathbf{c}$, (δ_c), $B(\cdot)$, $D(\cdot)$, $\mathbf{u}$, λ, f, l, x, R, t has the same meaning as the section above, and δ_c denotes possible motion direction, mdl denotes target model, $feature$ denotes feature extraction function adopted.

For a video sequence which contains tracking failures that result from model degradation, progressive updating is implemented to provide spatial and temporal information for progressive model updating, which can serve as a preprocess on data before applying the standard tracking process.

Algorithm 1 Tracking with progressive model updating

Input: video sequence f; target region $pos(0)$;
Output: tracking result pos
1: **for** $f_t(t > 1)$ **do**
2:     **for** each $B(\mathbf{c})$ **do**
3:         **for** each $\delta_{\mathbf{c}}$ **do**
4:             $D(\mathbf{c}) = D(f_t(\mathbf{c} + \delta_{\mathbf{c}}) - f_t(\mathbf{c}))$
5:         **end for**
6:         $\mathbf{u}_t(\mathbf{c}) = \delta_{\mathbf{c}_{\min}}$
7:     **end for**
8:     **for** each $\mathbf{u}_t(\mathbf{c})$ **do**
9:         $\mathbf{u}_\lambda(round(\mathbf{c} + \lambda\delta_{\mathbf{c}_{\min}})) = \mathbf{u}_t(\mathbf{c})$
10:     **end for**
11:     **for** each $\mathbf{u}_\lambda(\mathbf{c}_{\text{hole}})$ **do**
12:         $\mathbf{u}_\lambda(\mathbf{c}_{\text{hole}}) = Nearest(\mathbf{u}_\lambda(\mathbf{c}_{\text{assigned}}))$
13:     **end for**
14:     $f_\lambda(\mathbf{c}) = (1 - \lambda)f_{t-1}(\mathbf{c} - \mathbf{u}_\lambda(\mathbf{c})) + (\lambda) * f_t(\mathbf{c} + \mathbf{u}_\lambda(\mathbf{c}))$
15: **end for**
16: **for** $f_t(t > 1)$ **do**
17:     $x_{t-1} = feature(f_t, pos_{t-1})$
18:     $R = (mdl \cdot Corr(x_{t-1}, x_{t-1}))$;
19:     $pos_t = POS(R)$
20:     $x_t = feature(f_t, pos_t)$
21:     $mdl_t = \frac{label}{Corr(x_t, x_t)}$
22:     **if** $t = 1$ **then**
23:         $mdl = mdl_t$.
24:     **else**
25:         $mdl = (1 - l) \cdot mdl + l \cdot mdl_t$
26:     **end if**
27: **end for**

3.4 Effect of Progressive Updating

During model updating process, progressive model updating leads to a slightly higher model changing speed. Model updating with higher frequency compensates the insufficiency of small learning rate for fast motion targets, which implicitly leads to a tracking model updated in accordance to the changing of target.

During the detection process, in tracking failure cases, the peak of the response map is not as bright as functional tracking response map. Such ambiguity will usually result in failure to accurately recognize the true target. Progressive updating decelerates the training process for the tracker, and provides more iterations for the tracker to be updated. Higher updating frequency for the filter can lead to a clearer model for the target, which results in response maps with higher peak value and less ambiguity.

4 Experiment

4.1 Dataset

(1) VOT2017 dataset: The VOT2017 dataset [18] comprises 60 short sequences showing various objects in challenging backgrounds. The annotations are stored in a text file, and contains eight float number for each frame, which indicate the locations of the four corners of the bounding box.

(2) KITTI: The KITTI dataset [15] is a dataset for autonomous driving. The video sequence adopted is from the left color image of tracking dataset, and those from the static camera are training sequence 0016 and 0017. Other sequences are captured from a mobile camera on a vehicle, hence not adopted. Since the dataset is meant for multiple instances tracking, in order to make it suitable for single-object tracking methods, the data is transformed into tracking results with one object in ground truth each sequence. The regenerated ground truth sequence is stored in a text file with the top left coordinates, width, and height.

4.2 Methods of Evaluation

Since one baseline tracker [16] does not consider the scale-variance throughout the tracking process, the precision curve [4, 16, 24, 27] is adopted in order to evaluate the performance: the tracking result is considered precise when the distance between the predicted center and ground truth center is within the threshold. Precision curves show the percentage of correctly tracked frames for a range of distance thresholds [16]. A threshold of 20 pixels is chosen to demonstrate a representative precision score, as done in previous works [4, 16, 24, 27].

4.3 Comparison Scenarios and Experimental Details

The evaluation of the proposed method is over two experiments. The first experiment demonstrates performance improvement by implementing the method over the kernerlized correlation filter (KCF) tracker [16], and discriminative correlation filter with channel and spatial reliability(DCF-CSR) tracker [20] on VOT2017 dataset. The experiment result is also compared with most related trackers on VOT2017 challenge in Fig. 5, including SRDCF [7], Staple [3], and ECO [6] with handcrafted features. The second experiment is conducted on KITTI dataset with more challenging conditions with both trackers, and the experiment demonstrates an accuracy boost on both baseline trackers in these challenging scenarios.

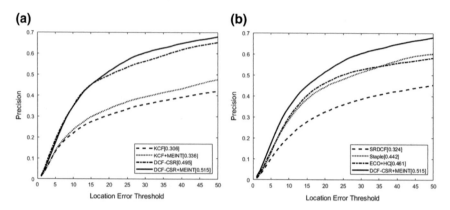

Fig. 5 The mean precision for 60 video sequences in VOT2017. The performance improvement of MEINT on KCF and DCF-CSR (left). Comparison between three recent most related trackers SRDCF, Staple, and ECO+HC (right)

During motion-estimated interpolation, the motion estimation is calculated with BMA [14], and the cost function is set as the pixel-wise sum of the absolute error. The hole area in the motion vector field is filled with the nearest calculated motion vector. The interpolation factor λ is set to 0.5. In the tracking module, 31-channel HOG features [5] using 4×4 cell size multiplied by a Hann window [4] is adopted. The learning (adaptation) rate of KCF and DCF-CSR $l = 0.02$ for most of the experiments, only fine-tuned to 0.0175 for KCF on KITTI single-tracklet dataset in order to get a slight performance improvement. The MATLAB implementation is tested on a machine equipped with an Intel Core i7 running at 3.40 GHz.

4.4 Experiment Result and Analysis

An ablation study on VOT2017 and KITTI was conducted to evaluate the contribution of motion-estimated progressive updating for KCF and CSR-DCF. Results of the precision plots are illustrated in Table 1.

Table 1 Precision score(20px) comparison between baseline model and the proposed method on VOT2017 and KITTI

	KCF		KCF+MEINT		DCF-CSR		DCF-CSR+MEINT	
Dataset	Precision	FPS	Precision	FPS	Precision	FPS	Precision	FPS
VOT2017	0.306	318.19	0.336	304.20	0.4954	10.2367	0.5150	10.5833
KITTI	0.429	334.98	0.435	311.30	0.408	10.01	0.4474	9.9936

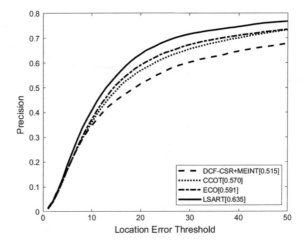

Fig. 6 The mean precision for 60 video sequences in VOT2017, compared with deep learning based trackers

As illustrated in Fig. 5 on dataset VOT2017, setting KCF tracking process on data without motion-estimated progressive updating (KCF) results in 8.9% performance drop in precision score compared to KCF+MEINT, and replacing the motion estimation interpolated data with raw data in DCF-CSR tracking process results in 3.8% drop in performance.

Figure 5b shows the precision score on the VOT2017. The proposed MEINT+ CSR-DCF outperforms correlation filter approaches with handcrafted features at the precision score 0.515.

For completeness purposes, recent deep learning-based trackers reported results on VOT2017 is included and illustrated in Fig. 6. The deep learning-based trackers still provides better results compared with handcrafted feature-based trackers due to better feature extraction mechanism with convolution neural network.

On KITTI dataset reformatted into single-object form, as illustrated in Fig. 7, the KCF tracking process with proposed method gain 0.5% performance improvement through learning rate fine-tuning from 0.02 to 0.0175, and MEINT improves the precision score of original DCF-CSR tracker by 9.7%.

Benefited from the additional information provided by progressive updating, the proposed method is significantly effective for tracking failure resulted from model degradation, which can be observed from Table 2. Due to additional information of consistency provided by interpolated frames, the progressive model updating assists the tracker to effectively adapt with the fast motion target. As illustrated in Fig. 3, due to illumination variance and fast camera motion, originally DCF-CSR tracker lost the target during a sharp illumination change, while the proposed method dealt with the situation well and led to a robust tracking result.

Compared with KCF tracker, the DCF-CSR method demonstrates higher performance improvement from MEINT. The reason is that the progressive updating encounters problems under occlusion scenarios, where background information occluding the target is responsible for the model degradation. However, the DCF-

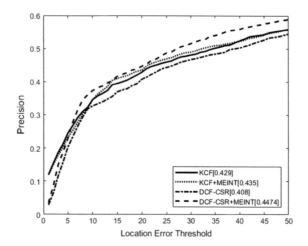

Fig. 7 Mean precision of single-object tracking with MEINT for 39 video sequences on KITTI

Table 2 Precision score improvement scenarios on VOT2017

Videos	KCF		KCF+MEINT		DCF-CSR		DCF-CSR+MEINT	
	Precision	FPS	Precision	FPS	Precision	FPS	Precision	FPS
Ball1	0.562	370.11	**0.600**	338.30	1.000	9.02	1.000	9.17
Birds1	0.009	545.80	0.009	554.40	0.322	8.56	**0.540**	7.83
bmx	0.145	244.16	**0.224**	226.18	0.105	7.13	**0.250**	4.82
Bolt1	0.683	504.85	**1.000**	471.19	1.000	13.6	1.000	13.5
Book	**0.029**	133.47	0.023	136.43	0.303	7.90	**0.526**	9.21
Drone1	0.037	780.70	0.037	738.73	0.076	9.92	**0.527**	10.3
Drone_across	0.048	333.15	**0.054**	299.81	0.245	9.33	**0.891**	9.76
Fish3	0.507	170.70	**0.805**	165.41	**0.611**	12.2	0.584	12.1
Gymnastics1	0.162	216.48	**0.243**	182.80	**0.996**	11.6	0.995	12.9
Handball2	0.353	310.40	**0.689**	290.74	0.381	10.90	**0.697**	11.04
Motocross1	0.085	269.74	**0.104**	238.97	0.122	10.8	**0.579**	11.6
Road	0.088	365.12	**0.998**	354.31	0.989	10.95	**0.993**	10.02
Shaking	0.019	202.24	**0.038**	195.63	0.888	11.4	**0.918**	12.1
Soccer1	0.263	135.25	**0.653**	108.76	0.429	11.35	**0.883**	11.59

CSR method contains a subsection which is designed to distinguish background from foreground [20], which makes the model training process robust against disturbance such as occlusion, leading to a more effective MEINT. As illustrated in Fig. 2, in video sequence 'soccer1' the tracking results remain stable against occlusion.

5 Conclusion

We propose to address the tracking model degradation problem of existing CF-based tracking methods via a progressive model updating mechanism based on motion-estimated interpolation. By progressive model updating using motion-estimated interpolation, intermediate response map can be obtained to fit the learning rates well, and thus learning-related model degradation can be alleviated effectively. Extensive experiments on two public tracking benchmarks and comparisons with recent state-of-the-art CF-based approaches demonstrate the effectiveness of the proposed method for fast motion object tracking.

Acknowledgements This work was partly funded by NSFC(NO.61571297), The National Key Research and Development Program (2017YFB1002401), and STCSM(18DZ2270700).

References

1. Babenko B, Yang MH, Belongie S (2011) Robust object tracking with online multiple instance learning. IEEE Trans Pattern Anal Mach Intell 33(8):1619–1632. https://doi.org/10.1109/TPAMI.2010.226
2. Baker S, Roth S, Scharstein D, Black MJ, Lewis JP, Szeliski R (2007) A database and evaluation methodology for optical flow. In: 2007 IEEE 11th international conference on computer vision, pp 1–8. https://doi.org/10.1109/ICCV.2007.4408903
3. Bertinetto L, Valmadre J, Golodetz S, Miksik O, Torr PHS (2016) Staple: complementary learners for real-time tracking. In: 2016 IEEE conference on computer vision and pattern recognition (CVPR), pp 1401–1409. https://doi.org/10.1109/CVPR.2016.156
4. Bolme DS, Beveridge JR, Draper BA, Lui YM (2010) Visual object tracking using adaptive correlation filters. In: 2010 IEEE computer society conference on computer vision and pattern recognition, pp 2544–2550. https://doi.org/10.1109/CVPR.2010.5539960
5. Dalal N, Triggs B (2005) Histograms of oriented gradients for human detection. In: Conference on computer vision and pattern recognition (CVPR)
6. Danelljan M, Bhat G, Khan FS, Felsberg M (2017) Eco: efficient convolution operators for tracking. In: 2017 IEEE conference on computer vision and pattern recognition (CVPR), pp 6931–6939. https://doi.org/10.1109/CVPR.2017.733
7. Danelljan M, Häger G, Khan FS, Felsberg M (2015) Learning spatially regularized correlation filters for visual tracking. In: 2015 IEEE international conference on computer vision (ICCV), pp 4310–4318. https://doi.org/10.1109/ICCV.2015.490
8. Danelljan M, Häger G, Khan FS, Felsberg M (2014) Accurate scale estimation for robust visual tracking. In: British machine vision conference, pp 65.1–65.11
9. Fan H, Ling H (2017) Sanet: structure-aware network for visual tracking. In: 2017 IEEE conference on computer vision and pattern recognition workshops (CVPRW), pp 2217–2224. https://doi.org/10.1109/CVPRW.2017.275
10. Galoogahi HK, Fagg A, Lucey S (2017) Learning background-aware correlation filters for visual tracking. In: 2017 IEEE international conference on computer vision (ICCV), pp 1144–1152. https://doi.org/10.1109/ICCV.2017.129
11. Galoogahi HK, Sim T, Lucey S (2013) Multi-channel correlation filters. In: 2013 IEEE international conference on computer vision, pp 3072–3079. https://doi.org/10.1109/ICCV.2013.381

12. Galoogahi HK, Sim T, Lucey S (2015) Correlation filters with limited boundaries. In: 2015 IEEE conference on computer vision and pattern recognition (CVPR), pp 4630–4638. https://doi.org/10.1109/CVPR.2015.7299094
13. Gao J, Ling H, Hu W, Xing J (2014) Transfer learning based visual tracking with gaussian processes regression. Springer International Publishing, New York
14. Gao XQ, Duanmu CJ, Zou CR (2000) A multilevel successive elimination algorithm for block matching motion estimation. IEEE Trans Image Process 9(3):501–504. https://doi.org/10.1109/83.826786
15. Geiger A, Lenz P, Urtasun R (2012) Are we ready for autonomous driving? the kitti vision benchmark suite. In: Proceedings of the conference on computer vision and pattern recognition (CVPR)
16. Henriques JF, Caseiro R, Martins P, Batista J (2015) High-speed tracking with kernelized correlation filters. IEEE Trans Pattern Anal Mach Intell 37(3):583–596. https://doi.org/10.1109/TPAMI.2014.2345390
17. Henriques JF, Carreira J, Rui C, Batista J (2014) Beyond hard negative mining: efficient detector learning via block-circulant decomposition. In: IEEE international conference on computer vision, pp 2760–2767
18. Kristan M, Matas J, Leonardis A, Vojir T, Pflugfelder R, Fernandez G, Nebehay G, Porikli F, Čehovin L (2016) A novel performance evaluation methodology for single-target trackers. IEEE Trans Pattern Anal Mach Intell 38(11):2137–2155. https://doi.org/10.1109/TPAMI.2016.2516982
19. Li R, Lv Y, Liu Z (2018) Multi-scheme frame rate up-conversion using space-time saliency. IEEE Access 6:1905–1915. https://doi.org/10.1109/ACCESS.2017.2780822
20. Lukežic A, Vojír T, Zajc LC, Matas J, Kristan M (2017) Discriminative correlation filter with channel and spatial reliability. In: 2017 IEEE conference on computer vision and pattern recognition (CVPR), pp. 4847–4856. https://doi.org/10.1109/CVPR.2017.515
21. Ma C, Yang X, Zhang C, Yang MH (2015) Long-term correlation tracking. In: 2015 IEEE conference on computer vision and pattern recognition (CVPR), pp 5388–5396. https://doi.org/10.1109/CVPR.2015.7299177
22. Nam H, Han B (2016) Learning multi-domain convolutional neural networks for visual tracking. In: 2016 IEEE conference on computer vision and pattern recognition (CVPR), pp 4293–4302. https://doi.org/10.1109/CVPR.2016.465
23. Revaud J, Weinzaepfel P, Harchaoui Z, Schmid C (2015) EpicFlow: edge-preserving interpolation of correspondences for optical flow. In: Proceedings of the conference on computer vision and pattern recognition
24. Rui C, Martins P, Batista J (2012) Exploiting the circulant structure of tracking-by-detection with kernels. In: European conference on computer vision, pp 702–715
25. Thang NV, Lee HJ (2017) An efficient non-selective adaptive motion compensated frame rate up conversion. In: 2017 IEEE international symposium on circuits and systems (ISCAS), pp 1–4. https://doi.org/10.1109/ISCAS.2017.8050462
26. Valmadre J, Bertinetto L, Henriques J, Vedaldi A, Torr PHS (2017) End-to-end representation learning for correlation filter based tracking. In: 2017 IEEE conference on computer vision and pattern recognition (CVPR), pp 5000–5008. https://doi.org/10.1109/CVPR.2017.531
27. Wu Y, Lim J, Yang MH (2013) Online object tracking: a benchmark. In: 2013 IEEE conference on computer vision and pattern recognition, pp 2411–2418. https://doi.org/10.1109/CVPR.2013.312

Index

A
Adversarial learning, 18–21, 24

C
Classification, 3–7, 17, 18, 20, 27, 28, 38, 51,
54, 57, 61, 63, 69, 81–83, 87, 89–91,
100–105, 112–115, 118, 119, 121–
124, 126
Correlation filter, 129, 130, 132, 133, 137,
139

D
Deep fusion, 97
Deep learning, 1, 2, 51–53, 68, 81, 84, 88,
89, 91, 96, 98, 112, 132, 139
Deep neural network, 81–83, 101
Domain adaptation, 1, 2, 4–11, 17–21, 25,
26, 29, 33, 35, 36, 45, 47, 81–84, 87–
92, 95
Domain discrepancy, 19

E
Ensemble learning, 65–67, 72, 77
Experimental, 20, 25, 35, 41, 48, 54, 55, 63,
92, 96, 108, 121, 137

F
Facial Action Unit Recognition, 95, 98–100,
105
Feature fine-tuning, 95
Feature fusion, 97, 107
Feature learning, 113, 132
Fine-tuning, 52, 54, 57, 59–63

G
Generative models, 9, 34

I
Image classification, 7, 18, 51, 54, 61, 63,
69, 81–83, 87, 89–91, 126
Intuition, 62, 98, 111–121, 123–126, 130

M
Metric learning, 17, 20, 29
Multi-modal representation learning, 99,
105

P
Progressive updating, 129, 130, 132–136,
138, 139

R
Reinforcement learning, 111, 113–115,
117–119, 123, 126

S
Style transfer, 33–36, 41, 42, 47

T
Transfer learning, 2–6, 10, 18, 20, 51–54,
56–58, 61, 63, 113, 114
Triplet loss, 17, 18, 20, 22, 27, 29, 73

U
Unsupervised classification, 18

© Springer Nature Switzerland AG 2020
R. Singh et al. (eds.), *Domain Adaptation for Visual Understanding*,
https://doi.org/10.1007/978-3-030-30671-7

Unsupervised domain adaptation, 17, 18, 20,
    25, 26, 29, 82–84, 87–91

V
Video segment retrieval, 65, 66, 69
Visual adaptation, 91
Visual tracking, 129

Printed in the United States
by Baker & Taylor Publisher Services